T0275754

LONDON MATHEMATICAL SOCIETY LECTURE NOTE SERIES

Managing Editor: Professor J.W.S. Cassels, Department of Pure Mathematics and Mathematical Statistics, University of Cambridge, 16 Mill Lane, Cambridge CB2 1SB, England

The titles below are available from booksellers, or, in case of difficulty, from Cambridge University Press.

London Mathematical Society Lecture Note Series. 209

Arithmetic of Diagonal Hypersurfaces over Finite Fields

Fernando Q. Gouvêa
Colby College, Maine

Noriko Yui
Queen's University, Kingston

CAMBRIDGE
UNIVERSITY PRESS

CAMBRIDGE UNIVERSITY PRESS
Cambridge, New York, Melbourne, Madrid, Cape Town, Singapore, São Paulo

Cambridge University Press
The Edinburgh Building, Cambridge CB2 2RU, UK

Published in the United States of America by Cambridge University Press, New York

www.cambridge.org
Information on this title: www.cambridge.org/9780521498340

First published 1995

A catalogue record for this publication is available from the British Library

Library of Congress Cataloguing in Publication data
Gouvêa, Fernando Q. (Fernando Quadros)
Arithmetic of diagonal hypersurfaces over finite fields / Fernando Q.Gouvêa
& Noriko Yui.
    p.   cm. - (London Mathematical Society lecture note series; 209)
ISBN 0 521 49834 1 (pbk.)
1. Hypersurfaces. 2. Finite fields (Algebra). I. Yui, Noriko.
II. Title. III. Series.
QA641.G67 1995
516.3'53-dc20 94-23790 CIP

ISBN-13 978-0-521-49834-0 paperback
ISBN-10 0-521-49834-1 paperback

Transferred to digital printing 2005

Contents

Acknowledgments

During part of the time of preparation of this work, F. Q. Gouvêa was on leave from the Universidade de São Paulo, Brazil and a visiting professor at Queen's University supported by the NSERC Individual Operating Grant and by the second author's ARC grant from Queen's University. During the remaining period, Gouvêa was at Colby College and was partially supported by NSF grants DMS–9203469 and DMS–9401313. The computations were conducted, for the most part, at Colby College, and were supported in part by a laboratory instrumentation grant from the College. Gouvêa would like to thank all of these institutions for their support.

During the course of this work, N. Yui was partially supported by an NSERC Individual Operating Grant and an NSERC Research Grant. During part of the time of preparation of this work, N. Yui held several fellowships at Newnham College and at the Department of Pure Mathematics and Mathematical Statistics (DPMMS), and the Newton Institute, University of Cambridge, supported by Newnham College, DPMMS, NSERC and the Royal Society of London. N. Yui is grateful to all members of DPMMS and the Newton Institute, and Fellows at Newnham College for their hospitality. N. Yui is especially indebted to J. Coates, R. Pinch and R. Taylor for their interest in this work and for fruitful discussions. These notes were finally completed at the Max–Planck–Institut für Mathematik, Bonn in the spring of 1994 while N. Yui held a visiting professorship supported by the MPIM Bonn. The hospitality and stimulating, friendly atmosphere of the institute are gratefully acknowledged.

Finally, we thank N. Boston, S. Kudla, S. Lichtenbaum, B. Mazur, N. Suwa, and D. Zagier for their interest in this work and for their comments and suggestions.

Notation and conventions

p: a rational prime number

$k = \mathbb{F}_q$: the finite field of q elements of $\mathrm{char}(k) = p > 0$

$k^\times = \langle z \rangle$: the multiplicative group of k with a fixed generator z

$\bar{k}$: the algebraic closure of k

$(k^\times)^m := \{c^m \mid c \in k^\times\}$

$\Gamma = \mathrm{Gal}(\bar{k}/k)$: the Galois group of $\bar{k}$ over k

$W = W(k)$: the ring of infinite Witt vectors over k

$K = K(k)$: the field of quotients of W

ν: a p-adic valuation of $\overline{\mathbb{Q}}_p$ normalized by $\nu(q) = 1$

F: the Frobenius morphism

V: the Verschiebung morphism

Φ: the Frobenius endomorphism

m and n: positive integers such that $m \geq 3, (m, p) = 1$ and $n \geq 1$

ℓ: a prime such that $(\ell, m) = 1$

$\mathbb{Q}_\ell$: the field of ℓ-adic rationals

$\mathbb{Z}_\ell$: the ring of ℓ-adic integers

$\mid \; \mid_\ell^{-1}$: the ℓ-adic valuation of $\mathbb{Q}$ normalized by $|\ell|_\ell^{-1} = \ell$

$|x|$: the absolute value of $x \in \mathbb{R}$

$L = \mathbb{Q}(\zeta)$: the m-th cyclotomic field over $\mathbb{Q}$ where $\zeta = e^{2\pi i/m}$

$G = \mathrm{Gal}(L/\mathbb{Q})$: the Galois group of L over $\mathbb{Q}$, which is isomorphic to $(\mathbb{Z}/m\mathbb{Z})^\times$

$\phi(m)$: the Euler function

$\mathbf{c} = (c_0, c_1, \ldots, c_{n+1}) \in \underbrace{k^\times \times \cdots \times k^\times}_{n+2 \text{ copies}}$: the twisting vector

$\mathcal{V} = \mathcal{V}_n^m(\mathbf{c})$: the diagonal hypersurface $\sum_{i=0}^{n+1} c_i X_i^m = 0 \subset \mathbb{P}_k^{n+1}$ with the twisting vector $\mathbf{c}$ of degree m and dimension n

$\mathcal{X} = \mathcal{V}_n^m(\mathbf{1})$: the Fermat variety $\sum_{i=0}^{n+1} X_i^m = 0 \subset \mathbb{P}_k^{n+1}$ of degree m and dimension n with the trivial twist $\mathbf{c} = \mathbf{1}$

$\boldsymbol{\mu}_m$: the group of m-th roots of unity in $\mathbb{C}$ (or in $\bar{k}$)

$\mathfrak{G} = \mathfrak{G}_n^m = \boldsymbol{\mu}_m^{n+2}/\Delta$: a subgroup of the automorphism group $\mathrm{Aut}(\mathcal{V})$ of $\mathcal{V}$

$\hat{\mathfrak{G}}$: the character group of $\mathfrak{G}$

$\mathfrak{A} = \mathfrak{A}_n^m$: the set of all characters $\mathbf{a} = (a_0, a_1, \ldots, a_{n+1}) \in \hat{\mathfrak{G}}$ such that

$$a_i \in \mathbb{Z}/m\mathbb{Z}, \qquad a_i \not\equiv 0 \pmod{m}, \qquad \text{and} \qquad \sum_{i=0}^{n+1} a_i \equiv 0 \pmod{m}.$$

For $\mathbf{a} = (a_0, a_1, \ldots, a_{n+1}) \in \mathfrak{A}_n^m$,

$\|\mathbf{a}\| = \sum_{i=0}^{n+1} \langle \frac{a_i}{m} \rangle - 1$ where $\langle x \rangle$ is the fractional part of $x \in \mathbb{Q}$

$p_\mathbf{a}$: the projector defined in Definition 3.1

$j(\mathbf{a})$: a Jacobi sum of dimension n and degree m

$\mathcal{J}(\mathbf{c}, \mathbf{a})$: a twisted Jacobi sum of dimension n and degree m

$\tilde{\mathbf{a}}$: an induced character in $\mathfrak{A}_{n+d}^m$ for some $d \geq 1$

$j(\tilde{\mathbf{a}})$: an induced Jacobi sum of an appropriate dimension and degree m

$\mathcal{J}(\tilde{\mathbf{c}}, \tilde{\mathbf{a}})$: an induced twisted Jacobi sum of an appropriate dimension and degree m

$A = [\mathbf{a}]$: the $(\mathbb{Z}/m\mathbb{Z})^\times$-orbit of $\mathbf{a}$

$p_A = [\mathbf{a}] = \sum_{\mathbf{a} \in A} p_\mathbf{a}$

$\tilde{A} = [\tilde{\mathbf{a}}]$: the $(\mathbb{Z}/m\mathbb{Z})^\times$-orbit of $\tilde{\mathbf{a}}$

$\mathcal{M}_A$: a Fermat motive of degree m and dimension n

$\mathcal{V}_A$: a twisted Fermat motive of degree m and dimension n

$\mathcal{M}_{\tilde{A}}$: an induced Fermat motive of degree m and an appropriate dimension

$\mathcal{V}_{\tilde{A}}$: an induced twisted Fermat motive of degree m and an appropriate dimension

$\#S$: the cardinality (resp. order) of a set (resp. group) S

$\mathfrak{B}_n^m = \{\mathbf{a} \in \mathfrak{A}_n^m \mid \mathfrak{J}(\mathbf{c}, \mathbf{a}) = q^{n/2}\}$ with n even

$\overline{\mathfrak{B}}_n^m = \{\mathbf{a} \in \mathfrak{A}_n^m \mid \mathfrak{J}(\mathbf{c}, \mathbf{a})/q^{n/2} = $ a root of unity in $L\}$ with n even

$\mathfrak{C}_n^m = \overline{\mathfrak{B}}_n^m \setminus \mathfrak{B}_n^m$

$\mathfrak{D}_n^m = \mathfrak{A}_n^m \setminus \mathfrak{B}_n^m$

$O(\mathfrak{C}_n^m)$: the set of $(\mathbb{Z}/m\mathbb{Z})^\times$-orbits in $\mathfrak{C}_n^m$

$O(\mathfrak{D}_n^m)$: the set of $(\mathbb{Z}/m\mathbb{Z})^\times$-orbits in $\mathfrak{D}_n^m$

$\varepsilon_d(\mathcal{V}_k) = \#O(\mathfrak{C}_n^m)$

$\lambda_d(\mathcal{V}_k) = \#O(\mathfrak{D}_n^m)$

$\delta_d(\mathcal{V}_k) = \varepsilon_d(\mathcal{V}_k) + \lambda_d(\mathcal{V}_k)$

Let M be a Γ-module where $\Gamma = \mathrm{Gal}(\bar{k}/k)$ with the Frobenius generator Φ.

M^Γ: the kernel of the map $\Phi - 1 : M \to M$

M_Γ: the cokernel of the map $\Phi - 1 : M \to M$

M_{tors}: the torsion subgroup of M

$\mathcal{O}$: the structure sheaf of $\mathcal{V}$ and $\mathcal{X}$

Ω: the sheaf of differentials on $\mathcal{V}$ and $\mathcal{X}$

$W\Omega$: the sheaf of de Rham–Witt complexes on $\mathcal{V}$ and $\mathcal{X}$

$\mathbb{G}_m$: the multiplicative group scheme

$\mathbb{G}_a$: the additive group scheme

Arithmetical invariants of $\mathcal{V}$ and $\mathcal{X}$ are rather sensitive to the fields of definition. Whenever the fields of definition are to be specified, subscripts are adjoined to the objects in question. For instance,

$\rho_r(\mathcal{V}_k)$ (resp. $\rho_r(\mathcal{V}_{\bar{k}})$): the r-th combinatorial Picard number of $\mathcal{V}$ defined over k (resp. $\bar{k}$)

$\rho'_r(\mathcal{V}_k)$ (resp. $\rho'_r(\mathcal{V}_{\bar{k}})$): the dimension of the subspace of $H^{2r}(\mathcal{V}_{\bar{k}}, \mathbb{Q}_\ell(r))$ generated by algebraic cycles of codimension r on $\mathcal{V}$ defined over k (resp. $\bar{k}$) where ℓ is a prime $\neq p$, which we call the r-th (geometric) Picard number of $\mathcal{V}$ defined over k (resp. $\bar{k}$)

$\mathrm{Br}^r(\mathcal{V}_k)$ (resp. $\mathrm{Br}^r(\mathcal{V}_{\bar{k}})$): the r-th "Brauer" group of $\mathcal{V}$ over k (resp. $\bar{k}$)

Introduction

Let $X = X_k$ be a smooth projective algebraic variety of dimension n defined over a finite field $k = \mathbb{F}_q$ of characteristic p. The zeta-function of X (relative to k) has the form

$$Z(X, q^{-s}) = \frac{P_1(X, q^{-s})P_3(X, q^{-s})\ldots P_{2n-1}(X, q^{-s})}{P_0(X, q^{-s})P_2(X, q^{-s})\ldots P_{2n}(X, q^{-s})}$$

where $P_i(X, T) \in 1 + T\mathbb{Z}[T]$ for every i, $0 \leq i \leq 2n$, and has reciprocal roots of absolute value $q^{i/2}$. Taking i equal to an even integer $2r$, we see that for any integer r between 0 and n

$$Z(X, q^{-s}) \sim \frac{C_X(r)}{(1 - q^{r-s})^{\rho_r(X)}} \quad \text{as } s \to r$$

where $C_X(r)$ is some rational number and $\rho_r(X)$ is an integer (called the r-th combinatorial Picard number of $X = X_k$). In this book, we obtain information about these two numbers for algebraic varieties that are especially simple.

There are standard conjectural descriptions of the numbers $\rho_r(X)$ and $C_X(r)$ that connect them with arithmetic and geometric invariants of X. Let $\bar{k}$ be an algebraic closure of k and let $X_{\bar{k}} := X \times_k \bar{k}$ be the base change of X from k to $\bar{k}$. Let ℓ be any prime different from $p = \text{char}(k)$. Let $\rho'_{r,\ell}(X)$ denote the dimension of the subspace of the ℓ-adic étale cohomology group $H^{2r}(X_{\bar{k}}, \mathbb{Q}_\ell(r))$, generated by algebraic cycles of codimension r on X defined over k, and let

$$\rho'_r(X) := \max_{\ell \neq p} \rho'_{r,\ell}(X).$$

(The numbers $\rho'_{r,\ell}(X)$ are in fact presumed to be independent of the choice of the prime ℓ.) We call $\rho'_r(X)$ the r-th Picard number of $X = X_k$. It is known that $\rho'_r(X) \leq \rho_r(X)$, and one conjectures that they are in fact equal:

CONJECTURE 0.1 (THE TATE CONJECTURE) *With the definitions above, we have*

$$\rho_r(X) = \rho'_r(X).$$

1

This is known to hold in a number of special cases (rational surfaces, Abelian surfaces, products of two curves, certain Fermat hypersurfaces, etc.)

Picard numbers are, of course, very sensitive to the field of definition. In various contexts we will want to compare the Picard number of a variety X over k to the Picard number of its base change to extensions of k. As one runs over bigger and bigger finite extensions of k, the combinatorial Picard number eventually stabilizes. We will refer to the latter number as the r-th (combinatorial) *stable* Picard number of X and denote it by $\bar{\rho}_r(X)$.

As for the rational number $C_X(r)$, a series of conjectures has been formulated by Lichtenbaum [Li84, Li87, Li90] and Milne [Mil86, Mil88] (see also Etesse [Et88]). (The conjectures concern the existence of "motivic cohomology" and in particular of certain complexes of étale sheaves $\mathbb{Z}(r)$.)

CONJECTURE 0.2 (THE LICHTENBAUM–MILNE CONJECTURE) *Assume that the complex $\mathbb{Z}(r)$ exists and that the Tate conjecture holds for $X = X_k$. Then*

$$C_X(r) = \pm\chi(X, \mathbb{Z}(r)) \cdot q^{\chi(X,\mathcal{O},r)}$$

where

$$\chi(X, \mathcal{O}, r) := r\chi(X, \mathcal{O}_X) - (r-1)\chi(X, \Omega_X^1) + \cdots \pm \chi(X, \Omega_X^{r-1})$$

and $\chi(X, \mathbb{Z}(r))$ is the Euler–Poincaré characteristic of the complex $\mathbb{Z}(r)$.

For surfaces, this formula is equivalent to the Artin–Tate formula, which is known to be true whenever the Tate conjecture holds. For higher dimensional varieties, the conjectural formula is known to hold only in some special cases. Therefore, providing examples related to this conjecture seems to be of considerable interest.

The purpose of these notes is to offer a testing ground for the Lichtenbaum–Milne conjecture for diagonal hypersurfaces, explicitly evaluating the special values of zeta-functions at integral arguments. This is done by passing to the twisted Fermat motives associated to such varieties. Our investigation is both theoretical and numerical; the results of our computations are recorded in Appendix A.

We now proceed to set up the case we want to investigate. Let m and n be integers such that $m \geq 3$, $(p, m) = 1$ and $n \geq 1$. Let $\mathbf{c} = (c_0, c_1, \cdots, c_{n+1})$ be a vector where $c_i \in k^\times$ for each $i = 0, 1, \ldots, n+1$, and let $\mathcal{V} = \mathcal{V}_n^m(\mathbf{c}) \subset \mathbb{P}_k^{n+1}$ denote the diagonal hypersurface of dimension n and of degree m defined over $k = \mathbb{F}_q$ given by the equation

$$c_0 X_0^m + c_1 X_1^m + \cdots + c_{n+1} X_{n+1}^m = 0. \qquad (*)$$

We denote by $\mathcal{X} := \mathcal{V}_n^m(\mathbf{1})$ the Fermat hypersurface of dimension n and of degree m defined by the equation $(*)$ with $\mathbf{c} = (1, 1 \cdots, 1) = \mathbf{1}$. We call the vector $\mathbf{c}$ a *twisting* vector. Note that the vector $\mathbf{c} = (c_0, c_1, \dots, c_{n+1})$ is only defined up to multiplication by a non-zero constant, and further, that changing any of the coefficients by an element in $k^\times$ which is an mth power gives an isomorphic variety. We will call two such choices for $\mathbf{c}$ equivalent. We will denote the set of all vectors $\mathbf{c} = (c_0, \dots, c_{n+1})$, considered up to equivalence, by $\mathcal{C}$.

Throughout the book, we impose the hypothesis that k contains all the m-th roots of unity, which is equivalent to the condition that $q \equiv 1 \pmod{m}$.

The diagonal hypersurface $\mathcal{V} = \mathcal{V}_n^m(\mathbf{c})$ is a complete intersection, and its cohomology groups are rather simple (cf. Deligne [De73], Suwa [Su93]). Its geometry and arithmetic are closely connected to those of the Fermat hypersurface, $\mathcal{X} = \mathcal{V}_n^m(\mathbf{1})$. In fact, the eigenvalues of the Frobenius endomorphism for $\mathcal{X}$ are Jacobi sums, and those for $\mathcal{V}$ are *twisted* Jacobi sums, that is, Jacobi sums multiplied by some m-th root of unity. Furthermore, the geometric and topological invariants of $\mathcal{V}$, such as the Betti numbers, the (i, j)-th Hodge numbers, the slopes and the dimensions and heights of formal groups are independent of the twisting vectors $\mathbf{c}$ for the defining equation for $\mathcal{V}$, and therefore coincide with the corresponding quantities for $\mathcal{X}$. By contrast, arithmetical invariants of $\mathcal{V}$ (that are sensitive to the fields of definition), such as the Picard number, the group of algebraic cycles, and the intersection matrix, differ from the corresponding quantities for $\mathcal{X}$. Relations between these arithmetical invariants of $\mathcal{V}$ and the corresponding invariants of $\mathcal{X}$ are one of our main themes.

To understand the arithmetic of a diagonal hypersurface $\mathcal{V} = \mathcal{V}_n^m(\mathbf{c})$ of dimension n and degree m with twist $\mathbf{c}$, we use the natural group action to associate to it a family of motives which correspond to a particularly natural decomposition of the cohomology of $\mathcal{V}$, which we call the *motivic decomposition*. We call these (not necessarily indecomposable) motives *twisted Fermat motives*, and the direct sum of these motives is the motive attached to $\mathcal{V}$ itself. The arithmetic of these motives "glues together" to form the arithmetic of $\mathcal{V}$.

Let $\mathcal{V}_A$ denote a twisted Fermat motive. We say that $\mathcal{V}_A$ is *supersingular* if the Newton polygon of $\mathcal{V}_A$ has a pure slope $n/2$; $\mathcal{V}_A$ is *ordinary* if the Newton polygon of $\mathcal{V}_A$ coincides with the Hodge polygon of $\mathcal{V}_A$; and $\mathcal{V}_A$ is *of Hodge–Witt type* if the Hodge–Witt cohomology group $H^{n-i}(\mathcal{V}_A, W\Omega^i)$ is of finite type for every i, $0 \le i \le n$. (If $\mathcal{V}_A$ is ordinary, then it is of Hodge–Witt type, but the converse is not true.) Then passing to diagonal hypersurfaces $\mathcal{V}$, we say that $\mathcal{V}$ is *supersingular, ordinary,* and *of Hodge–Witt type* if every twisted Fermat motive $\mathcal{V}_A$ is supersingular, ordinary, and of Hodge–Witt type, respectively. Note that these properties are not disjoint at the motivic level (that is, motives can be ordinary and supersingular at the same time).

The set of all diagonal hypersurfaces has a rather elaborate *inductive structure*, relating hypersurfaces of fixed degree and varying dimension. We focus on two types of these: the first relating hypersurfaces of dimension n and $n+2$, and the second relating hypersurfaces of dimensions $n+1$ and $n+2$. This inductive structure is independent of the twisting vectors of the defining equation for V. As before, the inductive structure can be considered at the motivic level, and the arithmetic and geometry of motives are closely related to those of their induced motives of higher dimension. Cohomological realizations of these structures shed light, for instance, on the Tate conjecture and on special values of (partial) zeta-functions. (For details, see Chapter 4 below.) This inductive structure also plays a major role in the work of Ran and Shioda on the Hodge conjecture for complex Fermat hypersurfaces (see [Ran81] and [Sh79a, Sh79b], for example).

For diagonal hypersurfaces $V = V_n^m(\mathbf{c})$ of odd dimension $n = 2d+1$, the Tate conjecture is trivially true (Milne [Mil86]). For diagonal hypersurfaces of dimension $n = 2$, the Tate conjecture can be proved for any twist $\mathbf{c}$ over k on the basis of the results of Tate [Ta65] and Shioda and Katsura [SK79] for Fermat surfaces $\mathcal{X}_2^m$ over k. We obtain the following result.

THEOREM 0.3
Let $V = V_n^m(\mathbf{c})$ be a diagonal hypersurface with twist $\mathbf{c}$ and let $\mathcal{X} = V_n^m(1)$ be the Fermat variety, both of degree m and dimension $n = 2d$ over $k = \mathbb{F}_q$. Let $\rho_d(V)$ and $\rho_d(\mathcal{X})$ denote the d-th combinatorial Picard number of V and $\mathcal{X}$, respectively, and let $\overline{\rho}_d(V)$ and $\overline{\rho}_d(\mathcal{X})$ be the corresponding stable combinatorial Picard numbers. Then the following assertions hold:

1. *The combinatorial stable Picard numbers are given by*

$$\overline{\rho}_d(V) = \overline{\rho}_d(\mathcal{X}) = 1 + \sum B_n(V_A)$$

   where the sum runs over all supersingular twisted Fermat motives V_A, and $B_n(V_A)$ denotes the n-th Betti number of V_A.

2. *Assume that m is prime, $m > 3$. Then*

$$\rho_d(\mathcal{X}_k) = \overline{\rho}_d(V).$$

   That is, the actual d-th combinatorial Picard number of $\mathcal{X}_k$ is stable.

3. *Assume that m is prime, $m > 3$. Then*

$$\rho_d(V_k) \leq \rho_d(\mathcal{X}_k).$$

   Furthermore, the following are equivalent:

   (a) *V_k and $\mathcal{X}_k$ are isomorphic*
   (b) *$\rho_d(V_k) = \rho_d(\mathcal{X}_k)$*

(c) **c** *is equivalent to the trivial twist* **1**.

Part 3 is false in general for composite m: for some values of m, one can find twists **c** such that $\rho_d(\mathcal{V}_k) > \rho_d(\mathcal{X}_k)$. One can also find non-trivial twists such that $\rho_d(\mathcal{V}_k) = \rho_d(\mathcal{X}_k)$. See section A.3.

Shioda [Sh82a] has obtained a closed formula for the stable Picard number for surfaces of prime degree: if $n = 2$, m is a prime, and $p \equiv 1$ (mod m) then:

$$\bar{\rho}_1(\mathcal{V}) = 1 + 3(m-1)(m-2).$$

Similar formulas hold for higher-dimensional hypersurfaces.

PROPOSITION 0.4
Using definitions and notation as above,

1. *when $n = 4$, m is prime, and $p \equiv 1$ (mod m),*

$$\bar{\rho}_2(\mathcal{V}) = 1 + 5(m-1)(3m^2 - 15m + 20),$$

2. *when $n = 6$, m is prime, and $p \equiv 1$ (mod m),*

$$\bar{\rho}_3(\mathcal{V}) = 1 + 5 \cdot 7(m-1)(3m^3 - 27m^2 + 86m - 95).$$

When m is prime and $p \equiv 1$ (mod m), Shioda's method allows such formulas to be computed for any specific even dimension. (See Appendix B for the details.) Similar methods allow one to get formulas that hold for more general degrees. Of course, these formulas only give the stable Picard numbers. When there is a non-trivial twist or when m is composite, determining the actual Picard number over $\mathbb{F}_q$ is much more delicate. We have computed many of these—see the tables in Appendix A.

Given a vector $\mathbf{c} = (c_0, c_1, \ldots, c_{n+1})$ and a character $\mathbf{a} = (a_0, a_1, \ldots a_{n+1})$, we define, in the usual way, $\mathbf{c}^{\mathbf{a}} := c_0^{a_0} c_1^{a_1} \cdots c_{n+1}^{a_{n+1}}$. We say **c** is *extreme* if we have $\mathbf{c}^{\mathbf{a}} \notin (k^{\times})^m$ for any $\mathbf{a} = (a_0, a_1, \ldots, a_{n+1}) \in \mathfrak{A}_n^m$ with $j(\mathbf{a}) = q^d$. One reason extreme twists are interesting is the following observation.

THEOREM 0.5
*Let $\mathcal{V} = \mathcal{V}_n^m(\mathbf{c})$ be a diagonal hypersurface of dimension $n = 2d$ and prime degree $m > 3$ over $k = \mathbb{F}_q$. Suppose that **c** is extreme. Then the Tate conjecture holds for $\mathcal{V}_k$, and we have*

$$\rho'_d(\mathcal{V}_k) = \rho_d(\mathcal{V}_k) = 1.$$

In the case of an extreme twist, one can also determine the intersection pairing on the (one dimensional!) image of the d-th Chow group in the cohomology.

For general diagonal hypersurfaces $\mathcal{V} = \mathcal{V}_m^n(\mathbf{c})$, we can use the results of Ran [Ran81], Shioda [Sh79a, Sh79b, Sh83b] and Tate [Ta65] to establish the validity of the Tate conjecture in the following cases.

PROPOSITION 0.6

Let $V = V_n^m(\mathbf{c})$ be a diagonal hypersurface of dimension $n = 2d > 2$ and degree $m > 3$ with twist $\mathbf{c}$ over a finite field $\mathbb{F}_q$ with $q = p^j \equiv 1 \pmod{m}$. Then the Tate conjecture holds in the following cases:

1. *m is prime, any n, and $p \equiv 1 \pmod{m}$.*

2. *$m \leq 20$, any n, and $p \equiv 1 \pmod{m}$.*

3. *$m = 21$, $n \leq 10$, and $p \equiv 1 \pmod{m}$.*

4. *m and n arbitrary and there exists j such that $p^j \equiv -1 \pmod{m}$ (equivalently, V is supersingular).*

Since diagonal hypersurfaces $V = V_n^m(\mathbf{c})$ are complete intersections, their zeta-functions have the form:

$$Z(V, T) = \frac{Q(V, T)^{(-1)^{n+1}}}{\prod_{i=0}^{n}(1 - q^i T)}.$$

In our case, $Q(V, T)$ is a polynomial of degree $\frac{m-1}{m}\{(m-1)^{n+1} + (-1)^{n+2}\}$ with integral coefficients, which factors over $\mathbb{C}$ as

$$Q(V, T) = \prod_{\mathbf{a} \in \mathfrak{A}_n^m} (1 - \mathcal{J}(\mathbf{c}, \mathbf{a})T)$$

where the product is taken over all twisted Jacobi sums, $\mathcal{J}(\mathbf{c}, \mathbf{a})$.

Studying the asymptotic behaviour of the zeta-function as $s \to r$ clearly boils down, then, to studying the asymptotic behaviour of the polynomial $Q(V, q^{-s})$ as s tends to r, $0 \leq r \leq n$. To do this, we first evaluate the polynomials $Q(V_A, q^{-r})$ corresponding to motives V_A as $s \to r$, and then glue together the motivic quantities to yield the following global results.

THEOREM 0.7

Let $V = V_n^m(\mathbf{c})$ be a diagonal hypersurface with twist $\mathbf{c}$ and let $X = V_n^m(\mathbf{1})$ be the Fermat variety, both of dimension n and degree m over $k = \mathbb{F}_q$.

(I) *Let $n = 2d$ be even, and assume that the degree m is prime, and that $m > 3$. Put $Q^*(V, T) = (1 - q^d T)Q(V, T)$. Define quantities $\varepsilon_d(V_k)$, $\delta_d(V_k)$ and $w_V(r)$, as follows:*

$$\varepsilon_d(V_k) = \frac{\rho_d(V_{\bar{k}}) - \rho_d(V_k)}{m - 1}, \quad \delta_d(V_k) = \frac{B_n(V) - \rho_d(V_k)}{m - 1},$$

and for any r, $0 \leq r \leq n$,

$$w_V(r) = \sum_{i=0}^{r}(r - i)h^{i,n-i}(V).$$

Then the following assertions hold for the limit

$$\lim_{s \to d} \frac{Q^*(\mathcal{V}, q^{-s})}{(1 - q^{d-s})^{\rho_d(\mathcal{V}_k)}}.$$

1. If $\mathcal{V}$ is supersingular (resp. strongly supersingular), then the limit is equal to $\pm m^{\varepsilon_d(\mathcal{V}_k)}$ (resp. equal to 1).

2. If $\mathcal{V}$ is of Hodge–Witt type, then the limit takes the following form:

$$\pm \frac{B^d(\mathcal{V}_k) m^{\delta_d(\mathcal{V}_k)}}{q^{w_V(d)}}.$$

Here $B^d(\mathcal{V}_k)$ is the global "Brauer number" of $\mathcal{V}_k$. It is a positive integer, and is a square up to powers of m.

If **c** is extreme, then $B^d(\mathcal{V}_k)$ is a square.

(II) Let $n = 2d + 1$ and $m > 3$ be prime. Then for any integer r, $0 \le r \le d$,

$$Q(\mathcal{V}, q^{-r}) = \frac{D^r(\mathcal{V}_k)}{q^{w_V(r)}}$$

where $D^r(\mathcal{V}_k)$ is a positive integer, and $D^r(\mathcal{V}_k) = D^{n-r}(\mathcal{V}_k)$.

Detailed accounts of Theorem 0.7 can be found in Chapters 6 and 7 below. The hypothesis of m being prime is not a subtle one, and is present mostly for technical reasons. One expects that there are similar (though perhaps more complicated) formulas for the cases of composite m. Our calculations are in general agreement with this expectation; see the comments in Chapter 9.

For diagonal hypersurfaces $\mathcal{V} = \mathcal{V}_2^m(\mathbf{c})$ of dimension $n = 2$ and degree $m > 3$ with twist **c** over $k = \mathbb{F}_q$, the Tate conjecture holds for $\mathcal{V}$ over k, so that $\mathcal{V}$ satisfies the Artin–Tate formula relative to k (cf. Milne [Mil75]). One of the motivations of the Lichtenbaum–Milne conjecture is to generalize the Artin–Tate formula to higher (even) dimensional varieties. For diagonal hypersurfaces $\mathcal{V} = \mathcal{V}_n^m(\mathbf{c})$ of dimension $n = 2d$ with twist **c** over $k = \mathbb{F}_q$, Lichtenbaum and Milne have shown that assuming the existence of complexes of étale sheaves $\mathbb{Z}(r)$ having certain properties yields the following formula:

THEOREM 0.8
*Assume the étale complexes $\mathbb{Z}(r)$ exist and satisfy the conditions in [Mil86, Mil88]. Let $\mathcal{V} = \mathcal{V}_n^m(\mathbf{c})$ be a diagonal hypersurface of dimension $n = 2d$ and (prime) degree $m > 3$ with twist **c** over $k = \mathbb{F}_q$. Assume that the cycle map $\mathrm{CH}^d(\mathcal{V}_k) \to H^n(\mathcal{V}_k, \mathbb{Z}(d))$ is surjective and that the Tate conjecture holds for $\mathcal{V}_k$. Then $\mathcal{V}_k$ satisfies the Lichtenbaum–Milne formula:*

$$\lim_{s \to d} \frac{Q^*(\mathcal{V}, q^{-s})}{(1 - q^{d-s})^{\rho_d(\mathcal{V}_k)}} = \pm \frac{\# \mathrm{Br}^d(\mathcal{V}_k) |\det A^d(\mathcal{V}_k)|}{q^{\alpha_d V(d)}}, \tag{†}$$

where $\mathrm{Br}^d(\mathcal{V}_k) = \#H^{n+1}(\mathcal{V}_k, \mathbb{Z}(d))$ is the "Brauer" group of $\mathcal{V}_k$ and $\#\mathrm{Br}^d(\mathcal{V}_k)$ is its order, $A^d(\mathcal{V}_k)$ is the image of the d-th Chow group $\mathrm{CH}^d(\mathcal{V}_k)$ in $H^n(\mathcal{V}_{\bar{k}}, \hat{\mathbb{Z}}(d))$, $\{D_i \mid i = 1, \cdots, \rho_d(\mathcal{V}_k)\}$ is a $\mathbb{Z}$-basis for $A^d(\mathcal{V}_k)$, $\det A^d(\mathcal{V}_k) = \det(D_i \cdot D_j)$ is the determinant of the intersection matrix on $A^d(\mathcal{V}_k)$, and $\alpha_\mathcal{V}(d) = s^{n+1}(d) - 2s^n(d) + w_\mathcal{V}(d)$ where $w_\mathcal{V}(d) = \sum_{i=0}^{d}(d-i)h^{i,n-i}(\mathcal{V})$ with $h^{i,j} = \dim_k H^j(\mathcal{V}, \Omega^i)$, and $s^i(d) = \dim \underline{H}^i(\mathcal{V}, \mathbb{Z}_p(d))$ (as a perfect group scheme).

For the definition of $\underline{H}$, see Milne [Mil86], p. 307.)

We refer to the formula in this theorem as the Lichtenbaum–Milne formula. It is known to hold for $d = 1$ or $d = 2$ whenever the Tate conjecture holds. When the Brauer group $\mathrm{Br}^d(\mathcal{V}_k)$ exists, its order is a square, and this gives us a handle on the (otherwise quite mysterious) value of this term in the formula.

Since we can get information about the special values directly from properties of twisted Jacobi sums, we can compare these results with those predicted by the Lichtenbaum–Milne formula.

THEOREM 0.9

The notation of Theorem 0.8 remains in force. Assume that m is prime (so the Tate conjecture holds), that the complexes $\mathbb{Z}(r)$ exist, and that the cycle map $\mathrm{CH}^d(\mathcal{V}_k) \to H^n(\mathcal{V}_k, \mathbb{Z}(d))$ is surjective (so the Lichtenbaum–Milne formula ($\dagger$) on the preceding page is valid). Then we have, for m prime:

(I) The following assertions hold:

1. *If $\mathcal{V}_k$ is supersingular, then*

$$\#\mathrm{Br}^d(\mathcal{V}_k)|\det A^d(\mathcal{V}_k)| = q^{\alpha_\mathcal{V}(d)} m^{\varepsilon_d(\mathcal{V}_k)}.$$

2. *If $\mathcal{V}_k$ is of Hodge–Witt type, then*

$$\#\mathrm{Br}^d(\mathcal{V}_k)|\det A^d(\mathcal{V}_k)| = B^d(\mathcal{V}_k) m^{\delta_d(\mathcal{V}_k)}.$$

(II) For each prime ℓ with $(\ell, m) = 1$, the following assertions hold:

1. *For a prime ℓ with $(\ell, mp) = 1$,*

$$\#\mathrm{Br}^d(\mathcal{V}_k)_{\ell-\mathrm{tors}} = \begin{cases} 1 & \text{if } \mathcal{V}_k \text{ is supersingular} \\ |B^d(\mathcal{V}_k)|_\ell^{-1} & \text{if } \mathcal{V}_k \text{ is of Hodge–Witt type} \end{cases}$$

   and

$$|\det A^d(\mathcal{V}_k) \otimes_{\mathbb{Z}} \mathbb{Z}_\ell| = 1.$$

2. *For the prime $p = \mathrm{char}(k)$, if $\mathcal{V}_k$ is of Hodge–Witt type, then*

$$\#\mathrm{Br}^d(\mathcal{V}_k)_{p-\mathrm{tors}} = |B^d(\mathcal{V}_k)|_p^{-1} \quad \text{and} \quad |\det A^d(\mathcal{V}_k) \otimes_{\mathbb{Z}} \mathbb{Z}_p| = 1.$$

(III) *The following divisibility assertions hold:*

1. *If $\mathcal{V}_k$ is strongly supersingular, then $\mathrm{Br}^d(\mathcal{V}_{\bar{k}})$ is a p-group, and $|\det A^d(\mathcal{V}_{\bar{k}})|$ divides a power of p.*

2. *If $\mathcal{V}_k$ is of Hodge–Witt type, then $\#\mathrm{Br}^d(\mathcal{V}_k)$ is a square (with the possible exception of the m-part), and $|\det A^d(\mathcal{V}_k)|$ divides a power of m.*

These results are similar to those announced by Suwa [Su91a, Su91b] for Fermat hypersurfaces.

For extreme twists **c**, we get a more satisfactory result. The Tate conjecture holds with $\rho_d(\mathcal{V}_k) = 1$, and one can determine the contribution of the intersection pairing explicitly.

THEOREM 0.10
Let $\mathcal{V} = \mathcal{V}_n^m(\mathbf{c})$ be a diagonal hypersurface of prime degree $m > 3$ and dimension $n = 2d$ with an extreme twist $\mathbf{c}$ over $k = \mathbb{F}_q$. Then $\mathrm{CH}^d(\mathcal{V}_k)$ is generated over $\mathbb{Q}$ by only one class of algebraic cycles and $|\det A^d(\mathcal{V}_k)| = m$. In this case, the Lichtenbaum–Milne formula (formula (†) on page 7) holds modulo the existence of $\mathrm{Br}^d(\mathcal{V}_k)$, in the sense that

$$\lim_{s \to d} \frac{Q^*(\mathcal{V}, q^{-s})}{(1 - q^{d-s})^{\rho_d(\mathcal{V}_k)}} = \pm \frac{m^{\varepsilon_d(\mathcal{V}_k)-1} B^d(\mathcal{V}_k) |\det A^d(\mathcal{V}_k)|}{q^{\alpha_d \mathcal{V}_k}}$$

where $\varepsilon_d(\mathcal{V}_k)$ is as in Theorem 0.7. The exponent $\varepsilon_d(\mathcal{V}_k) - 1$ is even, and the Brauer number $B^d(\mathcal{V}_k)$ is a square.

When it is defined, the actual order of the Brauer group relates to the Brauer number $B^d(\mathcal{V}_k)$ by the formula

$$\pm \#\mathrm{Br}^d(\mathcal{V}_k) = m^{\varepsilon_d(\mathcal{V}_k)-1} B^d(\mathcal{V}_k) = m^{\varepsilon_d(\mathcal{V}_k)-1} \prod_{\mathcal{V}_A} [B^d(\mathcal{V}_A)]$$

where the product is taken over all non-supersingular twisted Fermat motives $\mathcal{V}_A$. The motivic Brauer numbers $B^d(\mathcal{V}_A)$ are squares (including the m-part) for every $\mathcal{V}_A$ in the product.

Motives often seem to occur with even multiplicity. Therefore, the fact that $B^d(\mathcal{V}_A)$ is a square for every $\mathcal{V}_A$ is significant.

About the Lichtenbaum–Milne conjecture (Conjecture 0.2), we have:

THEOREM 0.11
Let $\mathcal{V} = \mathcal{V}_n^m(\mathbf{c})$ be a diagonal hypersurface of dimension n and prime degree $m > 3$ with twist $\mathbf{c}$ over $k = \mathbb{F}_q$. Assume the existence of the complexes $\mathbb{Z}(r)$ for r, $0 \le r \le n$ and the surjectivity of the cycle map $\mathrm{CH}^d(\mathcal{V}_k) \to H^n(\mathcal{V}_k, \mathbb{Z}(d))$.

(I) *Let $n = 2d$, and take an extreme twist $\mathbf{c}$. If $\mathcal{V}_k$ is of Hodge–Witt type, then the exponent of q in the residue $C_\mathcal{V}(d)$ of the Lichtenbaum–Milne formula*

(Conjecture 0.2) is correct, that is,

$$\chi(\mathcal{V}, \mathcal{O}, d) = w_{\mathcal{V}}(d) = \sum_{i=0}^{d}(d-i)h^{i,n-i}(\mathcal{V}).$$

Furthermore, $\chi(\mathcal{V}_k, \mathbb{Z}(r))$ is given by

$$\chi(\mathcal{V}_k, \mathbb{Z}(d)) = \frac{(-1)^d q^{-d(d+1)/2} \prod_{i=1}^{d}(q^i-1)^{-2}}{B^d(\mathcal{V}_k) \cdot m^{\delta_d(\mathcal{V}_k)}} \in \mathbb{Q}.$$

(II) *Let $n = 2d+1$. Then for any r, $0 \le r \le d$, $\mathcal{V}_k$ satisfies the Lichtenbaum–Milne formula. That is, for any r, $0 \le r \le d$, we have $\rho_r(\mathcal{V}_k) = 1$, and the exponent of q is correct, i.e., $\chi(\mathcal{V}, \mathcal{O}, r) = w_{\mathcal{V}}(r)$. Moreover, $\chi(\mathcal{V}_k, \mathbb{Z}(r))$ is given explicitly by*

$$\chi(\mathcal{V}_k, \mathbb{Z}(r)) = \frac{D^r(\mathcal{V}_k)}{(-1)^r q^{r(r+1)/2} \prod_{i=1}^{r}(q^i-1)^2 \cdot \prod_{j=1}^{2d-2r}(1-q^{r+j})} \in \mathbb{Q}.$$

Many of these results were previously announced in Gouvêa and Yui [GY92].

Our theoretical results were supplemented by extensive computations, which are described in Appendix A. A special focus of the computations was to try to understand how the Picard number and the special value of the zeta-function depend on the twist **c**. The results raise a number of interesting questions, some of which we discuss in Chapter 9.

1 Twisted Jacobi sums

Let $m \geq 3$ and $n \geq 1$ be integers. Let $k = \mathbb{F}_q$ be a finite field of characteristic $p > 0$ and let $\bar{k}$ denote its algebraic closure. Assume that k contains all the m-th roots of unity, which is equivalent to the condition that $q \equiv 1 \pmod{m}$. We fix a multiplicative character χ of k of exact order m by choosing a generator z of $k^\times$ and defining

$$\chi : k^\times = \langle z \rangle \to \boldsymbol{\mu}_m$$

by

$$\chi(z) = e^{2\pi i/m} := \zeta .$$

Let Δ denote the image of the diagonal inclusion $\boldsymbol{\mu}_m \hookrightarrow \boldsymbol{\mu}_m^{n+2}$, and let

$$\mathfrak{G} = \mathfrak{G}_n^m := \boldsymbol{\mu}_m^{n+2}/\Delta = \left\{ g = (\zeta_0, \zeta_1, \dots, \zeta_{n+1}) \in \boldsymbol{\mu}_m^{n+2} \right\}/\Delta$$

and let $\hat{\mathfrak{G}}$ be its character group.

Let $L = \mathbb{Q}(e^{2\pi i/m}) = \mathbb{Q}(\zeta)$ be the m-th cyclotomic field over $\mathbb{Q}$. Then $\hat{\mathfrak{G}}$ can be identified with the set

$$\hat{\mathfrak{G}} \simeq \left\{ \mathbf{a} = (a_0, a_1, \dots, a_{n+1}) \mid a_i \in \mathbb{Z}/m\mathbb{Z}, \ \sum_{i=0}^{n+1} a_i \equiv 0 \pmod{m} \right\}$$

under the pairing

$$\hat{\mathfrak{G}} \times \mathfrak{G} \to L : (\mathbf{a}, g) \to \mathbf{a}(g) = \prod_{i=0}^{n+1} \zeta_i^{a_i}.$$

Let $\mathfrak{A}_n^m$ be a subset of $\hat{\mathfrak{G}}$ defined by

$$\mathfrak{A}_n^m = \left\{ \mathbf{a} = (a_0, a_1, \dots a_{n+1}) \in \hat{\mathfrak{G}} \mid a_i \not\equiv 0 \pmod{m} \quad \text{for every } i \right\}.$$

If there is no ambiguity, we write $\mathfrak{A}$ for $\mathfrak{A}_n^m$ to make the notation lighter.

To each $\mathbf{a} = (a_0, a_1, \dots, a_{n+1}) \in \mathfrak{A}_n^m$, we will associate several objects that will later prove to be closely related to the geometry of our varieties. First, we define the length of $\mathbf{a}$ to be

$$\|\mathbf{a}\| = \sum_{j=0}^{n+1} \left\langle \frac{a_j}{m} \right\rangle - 1,$$

11

where $\langle x \rangle = x - [x]$ is the fractional part of $x \in \mathbb{R}$.

Now let $G = \mathrm{Gal}(L/\mathbb{Q})$; as usual, we identify G with $(\mathbb{Z}/m\mathbb{Z})^{\times}$. Let $H = \{\, p^i \bmod m \mid 0 \le i < f \,\}$ be the decomposition group of a prime ideal $\mathfrak{p}$ in L lying above p, with $\mathrm{Norm}_{L/\mathbb{Q}}(\mathfrak{p}) = p^f$. Let $G/H = \{\, s_1, s_2, \ldots, s_t \,\}$, so that $f \cdot t = \phi(m)$ (where ϕ is the Euler function). Then we define

$$A_H(\mathbf{a}) = \sum_{p^i \in H} \|p^i \mathbf{a}\| = \|\mathbf{a}\| + \|p\mathbf{a}\| + \cdots + \|p^{f-1}\mathbf{a}\|.$$

It is probably worth noting that if $p \equiv 1 \pmod{m}$, then $f = 1$ and $A_H(\mathbf{a}) = \|\mathbf{a}\|$. For practical reasons, many of our computations were done under this hypothesis. Finally, we define an element in the integral group ring of G,

$$\omega(\mathbf{a}) = \sum_{i=1}^{t} A_H(s_i \mathbf{a}) \cdot s_i \in \mathbb{Z}[G].$$

For much of the paper, we will fix a twisting vector $\mathbf{c} = (c_0, c_1, \ldots, c_{n+1})$ with $n+2$ components $c_i \in k^{\times}$. Consider a diagonal hypersurface $\mathcal{V} = \mathcal{V}_n^m(\mathbf{c}) \subset \mathbb{P}_k^{n+1}$ over k defined by the equation

$$c_0 X_0^m + c_1 X_1^m + \cdots + c_n X_n^m + c_{n+1} X_{n+1}^m = 0.$$

Note that there is a natural action of $\mathfrak{G}$ on $\mathcal{V}$.

It will occasionally be useful to consider the set of all possible vectors $\mathbf{c}$. It is clear that multiplying each c_i by an m-th power in $k^{\times}$ gives an isomorphic $\mathcal{V}$, and that multiplying all the c_i by a scalar will leave $\mathcal{V}$ unchanged. Hence, we will identify the set of all $\mathbf{c}$'s (considered up to equivalence) with the set

$$\mathcal{C} = \underbrace{k^{\times}/(k^{\times})^m \times k^{\times}/(k^{\times})^m \times \cdots \times k^{\times}/(k^{\times})^m}_{n+2 \text{ copies}} / \Delta,$$

where Δ is the diagonal inclusion of $k^{\times}/(k^{\times})^m$ in the product. We will say that $\mathbf{c}$ is *trivial* when it is equivalent to $(1, 1, \ldots, 1)$, i.e., when $\mathcal{V}$ is isomorphic to the Fermat hypersurface $\mathcal{X}$.

The set $\mathcal{C}$ clearly has a group structure. In fact, since the field k contains the m-th roots of unity, $\mathcal{C}$ is isomorphic to the group $\mathfrak{G}$ of automorphisms of $\mathcal{V}$; in particular, there is an action of $\mathbf{a} \in \mathfrak{A}$ on $\mathcal{C}$,

$$\mathcal{C} \times \mathfrak{A} \longrightarrow k^{\times}/(k^{\times})^m \cong \boldsymbol{\mu}_m$$

given by

$$(\mathbf{c}, \mathbf{a}) \mapsto \mathbf{c}^{\mathbf{a}} = c_0^{a_0} c_1^{a_1} \cdots c_{n+1}^{a_{n+1}}.$$

This map will be the main tool for comparing the arithmetic of $\mathcal{V}$ to that of the Fermat hypersurface $\mathcal{X}$.

DEFINITION 1.1 *The* twisted Jacobi sum of dimension n *and of degree* m *(relative to* q *and* χ*) associated to the character* $\mathbf{a} = (a_0, a_1, \ldots, a_{n+1}) \in \hat{\mathfrak{G}}$ *and the twisting vector* $\mathbf{c} = (c_0, c_1, \ldots, c_{n+1})$ *is*

$$\mathfrak{J}(\mathbf{c}, \mathbf{a}) = \mathfrak{J}(\mathbf{c}, \mathbf{a})_{q, \chi} = \bar{\chi}(c_0^{a_0} c_1^{a_1} \ldots c_{n+1}^{a_{n+1}}) j(\mathbf{a}) = \bar{\chi}(\mathbf{c}^{\mathbf{a}}) j(\mathbf{a})$$

where $j(\mathbf{a})$ *is the Jacobi sum of dimension* n *and of degree* m. *That is,*

$$j(\mathbf{a}) = (-1)^n \sum \chi(v_1)^{a_1} \chi(v_2)^{a_2} \ldots \chi(v_{n+1})^{a_{n+1}}$$

where the sum is taken over all $(n+1)$*-tuples* $(v_1, v_2, \ldots, v_{n+1}) \in k^\times \times \cdots \times k^\times$ *subject to the linear relation* $1 + v_1 + \cdots + v_{n+1} = 0$. *We will often refer to the vector* $\mathbf{c}$ *or to the root of unity* $\bar{\chi}(c_0^{a_0} c_1^{a_1} \ldots c_{n+1}^{a_{n+1}})$ *as "the twist."*

PROPOSITION 1.2 (PROPERTIES OF TWISTED JACOBI SUMS)
Let $\mathfrak{J}(\mathbf{c}, \mathbf{a})$ *be a twisted Jacobi sum of dimension* n *and of degree* m. *Then* $\mathfrak{J}(\mathbf{c}, \mathbf{a})$ *has the following properties:*

1. $\mathfrak{J}(\mathbf{c}, \mathbf{a})$ *is an algebraic integer in* L: *more precisely, if* $\gcd(\mathbf{a}, m) = d \geq 1$, *then* $\mathfrak{J}(\mathbf{c}, \mathbf{a})$ *is an algebraic integer in* $\mathbb{Q}(\zeta^d) \subsetneq L$. *With respect to any complex embedding,* $\mathfrak{J}(\mathbf{c}, \mathbf{a})$ *has the absolute value* $|\mathfrak{J}(\mathbf{c}, \mathbf{a})| = q^{n/2}$.

2. *Let*
$$G(\chi) = \sum_{x \in k} \chi(x) \psi(x)$$
denote a Gauss sum, where $\psi : k \mapsto \mu_p$ *is the additive character of* k *defined by*
$$\psi(x) = e^{2\pi i(x + \cdots + x^{q/p})/p}.$$
Then for $\mathbf{a} = (a_0, a_1, \ldots, a_{n+1}) \in \mathfrak{A}_n^m$,
$$\mathfrak{J}(\mathbf{c}, \mathbf{a}) = \frac{1}{q} \bar{\chi}(c_0^{a_0} c_1^{a_1} \ldots c_{n+1}^{a_{n+1}}) G(\chi^{a_0}) G(\chi^{a_1}) \ldots G(\chi^{a_{n+1}}).$$

3. *As above, let*
$$G = \mathrm{Gal}(L/\mathbb{Q}) = \{\, \sigma_t \mid \sigma_t(\zeta) = \zeta^t \text{ with } t \bmod m \,\}$$
$$\cong \{ t \bmod m \mid (t, m) = 1 \}.$$
Then G *acts on* $\mathfrak{J}(\mathbf{c}, \mathbf{a})$ *by*
$$\mathfrak{J}(\mathbf{c}, \mathbf{a})^{\sigma_t} = \mathfrak{J}(\mathbf{c}, t\mathbf{a}) = \bar{\chi}(c_0^{ta_0} c_1^{ta_1} \ldots c_{n+1}^{ta_{n+1}}) j(t\mathbf{a}),$$
where of course
$$t\mathbf{a} = (ta_0, ta_1, \ldots, ta_{n+1}) \in \mathfrak{A}_n^m.$$
In other words, G *acts on* $\mathfrak{J}(\mathbf{c}, \mathbf{a})$ *via its natural action on* $\mathfrak{A}_n^m$.

4. Let $\mathfrak{p} \subset L = \mathbb{Q}(\zeta)$ be any ideal lying above p. As an ideal in L, $(\mathfrak{J}(\mathbf{c}, \mathbf{a}))$ has the prime ideal decomposition

$$(\mathfrak{J}(\mathbf{c}, \mathbf{a})) = \mathfrak{p}^{\omega(\mathbf{a})},$$

where $\omega(\mathbf{a}) \in \mathbb{Z}[G]$ is the element defined above.

5. For any prime ℓ with $(\ell, mp) = 1$, let $|\quad|_\ell$ denote the ℓ-adic absolute value normalized by $|\ell|_\ell^{-1} = \ell$. Then

$$|\mathfrak{J}(\mathbf{c}, \mathbf{a})|_\ell = |j(\mathbf{a})|_\ell.$$

6. Let ν denote the p-adic valuation of $\bar{\mathbb{Q}}_p$ normalized by $\nu(q) = 1$. Then

$$\nu(\mathfrak{J}(\mathbf{c}, \mathbf{a})) = \nu(j(\mathbf{a})).$$

Proof: The twisted Jacobi sum $\mathfrak{J}(\mathbf{c}, \mathbf{a})$ differs from the Jacobi sums $j(\mathbf{a})$ only by multiplication by an m-th root of unity. Thus, all assertions are clear. $\square$

There is an inductive structure in the twisted Jacobi sums which reflects an underlying geometric inductive structure (studied extensively by Shioda [Sh79a, Sh82b] for Fermat varieties). Fix a positive integer $m \geq 3$ and let r, s be positive integers. Define a set

$$\mathfrak{A}_{r,s}^m := \{(\mathbf{b}, \mathbf{d}) \in \mathfrak{A}_r^m \times \mathfrak{A}_s^m \mid \mathbf{b} = (b_0, b_1, \ldots, b_{r+1}), \mathbf{d} = (d_0, d_1, \ldots, d_{s+1})$$

$$\text{with} \quad b_{r+1} + d_{s+1} \equiv 0 \pmod{m}\}.$$

Then a pair $(\mathbf{b}, \mathbf{d}) \in \mathfrak{A}_{r,s}^m$ gives rise to a character in $\mathfrak{A}_{r+s}^m$:

$$\mathbf{b} \# \mathbf{d} := (b_0, b_1, \ldots, b_r, d_0, d_1, \ldots, d_s) \in \mathfrak{A}_{r+s}^m.$$

On the other hand, a pair $(\mathbf{b}', \mathbf{d}') \in \mathfrak{A}_{r-1}^m \times \mathfrak{A}_{s-1}^m$ also yields a character in $\mathfrak{A}_{r+s}^m$:

$$\mathbf{b}' * \mathbf{d}' := (b_0', b_1', \ldots, b_r', d_0', d_1', \ldots, d_s') \in \mathfrak{A}_{r+s}^m.$$

We are often interested in considering characters in $\mathfrak{A}_n^m$ up to permutation of the entries. From that point of view, these two inductive structures are clearly related: if $(\mathbf{b}, \mathbf{d}) \in \mathfrak{A}_{r,s}^m$ and we define $\mathbf{e} = (b_{r+1}, d_{s+1}) \in \mathfrak{A}_0^m$, then we have $\mathbf{a} = \mathbf{b} \# \mathbf{d}$ if and only if $\mathbf{a} * \mathbf{e} \sim \mathbf{b} * \mathbf{d}$, where $\sim$ denotes equality up to permutation of the entries.

Shioda [Sh79a] has shown that there is a bijection

$$\mathfrak{A}_{r+s}^m \longleftrightarrow \mathfrak{A}_{r,s}^m \cup (\mathfrak{A}_{r-1}^m \times \mathfrak{A}_{s-1}^m).$$

In other words, every Jacobi sum of dimension $m = r + s$ can be obtained from Jacobi sums of lower dimension by one of the two methods.

This inductive structure can be realized cohomologically (see Ran [Ran81] and Shioda [Sh79a]), and this is the basis of the work of Ran and Shioda on the Hodge conjecture for complex Fermat hypersurfaces. Here we shall discuss how the inductive structure is realized at the level of Jacobi sums and twisted Jacobi sums.

LEMMA 1.3
For fixed $m \geq 3$ and $n \geq 1$, choose r and s so that $r + s = n$, so that we have a bijection

$$\mathfrak{A}_n^m \longleftrightarrow \mathfrak{A}_{r,s}^m \cup \mathfrak{A}_{r-1}^m \times \mathfrak{A}_{s-1}^m.$$

1. *If $\mathbf{a} = \mathbf{b}\#\mathbf{d} = (b_0, b_1, \ldots, b_r, d_0, d_1, \ldots, d_s)$ with $(\mathbf{b}, \mathbf{d}) \in \mathfrak{A}_{r,s}^m$, then the Jacobi sum $j(\mathbf{a})$ of degree m and dimension n is given by*

$$j(\mathbf{a}) = \chi(-1)j(\mathbf{b})j(\mathbf{d}).$$

2. *If $\mathbf{a} = \mathbf{b}' * \mathbf{d}' = (b_0', b_1', \ldots, b_r', d_0, d_1, \ldots, d_s)$ with $(\mathbf{b}', \mathbf{d}') \in \mathfrak{A}_{r-1}^m \times \mathfrak{A}_{s-1}^m$, then the Jacobi sum $j(\mathbf{a})$ of degree m and dimension n is given by*

$$j(\mathbf{a}) = qj(\mathbf{b}')j(\mathbf{d}').$$

Proof: This is an immediate consequence of the product expression for $j(\mathbf{a})$ in terms of Gauss sums, and the identity $G(\chi)G(\bar{\chi}) = q\chi(-1)$.

For the first statement, we have

$$j(\mathbf{a}) = \frac{1}{q}G(\chi^{b_0})\ldots G(\chi^{b_r}) \cdot G(\chi^{d_0})\ldots G(\chi^{d_s})$$
$$= \frac{1}{q}[qj(\mathbf{b}) \cdot G(\chi^{b_{r+1}})^{-1}][qj(\mathbf{d})G(\chi^{d_{s+1}})^{-1}] = \chi(-1)j(\mathbf{b})j(\mathbf{d}),$$

because of the condition $b_{r+1} + d_{s+1} \equiv 0 \pmod{m}$.

As to the second statement,

$$j(\mathbf{a}) = \frac{1}{q}G(\chi^{b_0'})\ldots G(\chi^{b_r'})G(\chi^{d_0})\ldots G(\chi^{d_s}) = qj(\mathbf{b}')j(\mathbf{d}'),$$

as desired. $\square$

We will be particularly interested in the case when $r = n - 1$, $s = 1$, which will allow us to obtain information on Jacobi sums of dimension n by putting together information from dimensions $n - 1$, $n - 2$, 1, and 0.[1] Notice that if $\mathbf{d} \in \mathfrak{A}_0^m$, then $j(\mathbf{d}) = \chi(-1)$, which makes the formula in item 2 above particularly simple. It is probably also worth pointing out that when m is odd we must have $\chi(-1) = 1$, simplifying the formulas still further.

---

[1]The main drawback to this point of view appears when we want to concentrate on hypersurfaces of even dimension, since this forces us to relate them with hypersurfaces of odd dimension.

REMARK 1.4 Going down the inductive structure, we see that any character $\mathbf{a} \in \mathfrak{A}_n^m$ can be expressed in terms of characters in $\mathfrak{A}_0^m$ (which are trivial to understand), $\mathfrak{A}_1^m$ and $\mathfrak{A}_{1,1}^m$.

There is another (much simpler, but still useful) inductive structure on Jacobi sums which depends on the degree.

LEMMA 1.5
Fix $n \geq 1$, and let $m = m_0^t$ be a power of a prime m_0, and assume that either $m_0 \geq 3$ and $t \geq 2$, or $m_0 = 3$, $t \geq 3$. Let $\mathbf{a} = (a_0, a_1, \ldots, a_{n+1}) \in \mathfrak{A}_n^m$. Assume that $\gcd(a_0, a_1, \ldots, a_{n+1}, m) > 1$, and hence is a power of m_0. Write $a_i/m_0 = a_i'$ for each i, $0 \leq i \leq n+1$. Then $\mathbf{a}' = (a_0', a_1', \ldots, a_{n+1}')$ is an element of $\mathfrak{A}_n^{m_0^{t-1}}$. If we write $j_m(\mathbf{a})$ for the Jacobi sum of an element of $\mathfrak{A}_n^m$, then we have

$$j_{m_0^t}(\mathbf{a}) = j_{m_0^{t-1}}(\mathbf{a}').$$

Proof: This follows from the identity on Gauss sums:

$$G_{m_0^t}(\chi^{a_i}) = G_{m_0^{t-1}}(\chi^{a_i'}). \quad \square$$

Adding in the twist does not change much:

PROPOSITION 1.6
The inductive structures above are realized at the level of twisted Jacobi sums as follows:

(I) Fix $m \geq 3$ and vary $n \geq 1$.

1. Choose positive integers r and s such that $r + s = n$. Let

$$\mathbf{c} = (c_0, c_1, \ldots, c_{r+1}) \in \underbrace{k^\times \times \cdots \times k^\times}_{(r+2) \text{ copies}}$$

and

$$\mathbf{d} = (d_0, d_1, \ldots, d_{s+1}) \in \underbrace{k^\times \times \cdots \times k^\times}_{(s+2) \text{ copies}}.$$

Write $\tilde{\mathbf{c}} = (c_0, c_1, \ldots, c_r, d_0, d_1, \ldots, d_s)$.

Let $\mathbf{a} = (a_0, a_1, \ldots, a_{r+1}) \in \mathfrak{A}_r^m$ and $\mathbf{b} = (b_0, b_1, \ldots, b_{s+1}) \in \mathfrak{A}_s^m$ such that $a_{r+1} + b_{s+1} = m$. Let $\tilde{\mathbf{a}} := \mathbf{a} \# \mathbf{b} = (a_0, a_1, \ldots, a_r, b_0, b_1, \ldots, b_s) \in \mathfrak{A}_n^m$ be the induced character. Then

$$\mathfrak{J}(\tilde{\mathbf{c}}, \tilde{\mathbf{a}}) = \bar{\chi}(-d_{s+1}/c_{r+1})^{a_{r+1}} \mathfrak{J}(\mathbf{c}, \mathbf{a}) \mathfrak{J}(\mathbf{d}, \mathbf{b}).$$

2. *Choose positive integers r and s such that $r + s = n$. Let*

$$\mathbf{c} = (c_0, c_1, \ldots, c_{r+1}) \in \underbrace{k^\times \times \cdots \times k^\times}_{(r+1) \ copies}$$

and

$$\mathbf{d} = (d_0, d_1, \ldots, d_{s+1}) \in \underbrace{k^\times \times \cdots \times k^\times}_{(s+1) \ copies}.$$

Put $\tilde{\mathbf{c}} = (c_0, c_1, \ldots, c_r, d_0, d_1, \ldots, d_s)$.

Let $\mathbf{a} = (a_0, a_1, \ldots, a_{r+1}) \in \mathfrak{A}^m_{r-1}$, $\mathbf{b} = (b_0, b_1, \ldots, b_s) \in \mathfrak{A}^m_{s-1}$, and let $\tilde{\mathbf{a}} = (a_0, a_1, \ldots, a_r, b_0, b_1, \ldots, b_s) \in \mathfrak{A}^m_n$ be the induced character. Then we have

$$\mathfrak{J}(\tilde{\mathbf{c}}, \tilde{\mathbf{a}}) = q \mathfrak{J}(\mathbf{c}, \mathbf{a}) \mathfrak{J}(\mathbf{d}, \mathbf{b}).$$

3. *In particular, if we take $r = n - 1$, $s = 1$, let*

$$\mathbf{c} = (c_0, c_1, \ldots, c_{n-1}) \in \underbrace{k^\times \times \cdots \times k^\times}_{n \ copies}$$

be a twisting vector, and let $\mathbf{a} = (a_0, a_1, \ldots, a_{n-1}) \in \mathfrak{A}^m_{n-2}$. For any $a \in \mathbb{Z}/m\mathbb{Z}$, we get an induced character $\tilde{\mathbf{a}} = (a_0, a_1, \ldots, a_{n-1}, a, m-a) \in \mathfrak{A}^m_n$. For any pair (c_n, c_{n+1}) in $k^\times \times k^\times$, construct a new twisting vector $\tilde{\mathbf{c}} = (c_0, c_1, \ldots, c_{n-1}, c_n, c_{n+1})$. Then we have

$$\mathfrak{J}(\tilde{\mathbf{c}}, \tilde{\mathbf{a}}) = q \bar{\chi}(-c_n^a c_{n+1}^{m-a}) \mathfrak{J}(\mathbf{c}, \mathbf{a}) = q \bar{\chi}(-c_n/c_{n+1})^a \mathfrak{J}(\mathbf{c}, \mathbf{a}).$$

Consequently, for any integer r, $0 \le r \le n$,

$$\mathrm{Norm}_{L/\mathbb{Q}} \left(1 - \frac{\mathfrak{J}(\mathbf{c}, \mathbf{a})}{q^r} \right) = \mathrm{Norm}_{L/\mathbb{Q}} \left(1 - \bar{\chi}(-c_{n+2}/c_{n+3})^a \frac{\mathfrak{J}(\tilde{\mathbf{c}}, \tilde{\mathbf{a}})}{q^{r+1}} \right).$$

(II) *Fix n and vary m. Suppose that $m = m_0^t$ where m_0 is a prime and either $m > 3$ and $t \ge 2$, or $m_0 = 3$ and $t \ge 3$. Let*

$$\mathbf{c} = (c_0, c_1, \ldots, c_{n+1}) \in \underbrace{k^\times \times \cdots \times k^\times}_{(n+2) \ copies}.$$

Let $\mathbf{a} = (a_0, a_1, \ldots, a_{n+1}) \in \mathfrak{A}^m_n$ such that $\gcd(a_0, a_1, \ldots, a_{n+1}, m) \ne 1$. Put $m' = m/m_0 = m_0^{t-1}$ and let $\mathbf{a}' = (a'_0, a'_1, \ldots, a'_{n+1}) \in \mathfrak{A}^{m'}_n$ where $a'_i = a_i/m_0$ for each i. Then

$$\mathrm{Norm}_{\mathbb{Q}(\zeta_m)/\mathbb{Q}}(1 - \frac{\mathfrak{J}(\mathbf{c}, \mathbf{a})}{q^r}) = \mathrm{Norm}_{\mathbb{Q}(\zeta_{m'})/\mathbb{Q}}(1 - \frac{\mathfrak{J}(\mathbf{c}, \mathbf{a}')}{q^r})$$

for any integer r, $0 \le r \le n$.

Proof: We consider the first statement in part (I). From Lemma 1.3, part 1, we have

$$\mathcal{J}(\tilde{\mathbf{c}}, \tilde{\mathbf{a}}) = \bar{\chi}(c_0^{a_0} c_1^{a_1} \ldots c_r^{a_r} d_0^{b_0} d_1^{b_1} \ldots d_s^{b_r}) j(\mathbf{a} \# \mathbf{b})$$
$$= \bar{\chi}(c_0^{a_0} c_1^{a_1} \ldots c_r^{a_r} c_{r+1}^{a_{r+1}}) \bar{\chi}(d_0^{b_0} d_1^{b_1} \ldots b_s^{b_s} d_{s+1}^{b_{s+1}}) \bar{\chi}(c_{r+1}^{-a_{r+1}} d_{s+1}^{-b_{s+1}}) j(\mathbf{a} \# \mathbf{b})$$
$$= \bar{\chi}(-d_{s+1}/c_{r+1})^{a_{r+1}} \mathcal{J}(\mathbf{c}, \mathbf{a}) \mathcal{J}(\mathbf{d}, \mathbf{b}).$$

The second and third statements are proved in the same way.

For part (II), let χ_m and $\chi_{m'}$ be multiplicative characters of $k^\times$ of exact order m and m', respectively. Just note that

$$\chi_m(c_0^{a_0} c_1^{a_1} \ldots c_{n+1}^{a_{n+1}}) = \chi_m((c_0^{a_0'} c_1^{a_1'} \ldots c_{n+1}^{a_{n+1}'})^{m_0})$$
$$= \chi_{m'}(c_0^{a_0'} c_1^{a_1'} \ldots c_{n+1}^{a_{n+1}'}).$$

This together with Lemma 1.5 then yields the assertion. □

In what follows, this inductive structure will have to be dealt with on two levels: one may look only at the characters $\mathbf{a}$, or consider also the twists $\mathbf{c}$. In the first case, as we shall see, one obtains information about geometric properties of the diagonal hypersurface; taking the twisting vector $\mathbf{c}$ into account is only necessary when looking for arithmetic properties. It is clear from the results above that the inductive structure works much better when we can ignore the twist $\mathbf{c}$.

PROPOSITION 1.7

If m is an odd prime power, then the twisted Jacobi sum $\mathcal{J}(\mathbf{c}, \mathbf{a})$ satisfies the congruence

$$\mathcal{J}(\mathbf{c}, \mathbf{a}) \equiv 1 \pmod{(1 - \zeta)},$$

where ζ is an m-th root of unity in L. If $m > 3$ and $\mathbf{c}^{\mathbf{a}}$ is an m-th power in k, then in fact

$$\mathcal{J}(\mathbf{c}, \mathbf{a}) = j(\mathbf{a}) \equiv 1 \pmod{(1 - \zeta)^3}.$$

Proof: The twisted Jacobi sum $\mathcal{J}(\mathbf{c}, \mathbf{a})$ differs from the Jacobi sums $j(\mathbf{a})$ only by multiplication by an m-th root of unity. If m is prime, then the Jacobi sum $j(\mathbf{a})$ satisfies the congruence

$$j(\mathbf{a}) \equiv 1 \pmod{(1 - \zeta)^\alpha} \quad \text{with} \quad \alpha \in \mathbb{Z}, \quad \alpha \geq 2.$$

In fact, if $m > 3$ it is known that $\alpha \geq 3$. (This is due to Iwasawa [Iw75].) The Iwasawa congruence can be generalized, using a result of Ihara that covers the case $n = 1$, to prime powers. This fact was stated in Shioda [Sh87] without proof. We shall include a short proof here invoking the inductive structures, and the Ihara congruence for Jacobi sums of dimension 1.

If $m = m_0^t$ with m_0 prime and $m > 3$, then Ihara [Ih86] has shown that a Jacobi sum $j(\mathbf{a})$ of dimension one satisfies the congruence

$$j(\mathbf{a}) \equiv 1 \pmod{(\zeta^{a_0} - 1)(\zeta^{a_1} - 1)(\zeta^{a_2} - 1)}$$

where $\mathbf{a} = (a_0, a_1, a_2) \in \mathfrak{A}_1^m$ with $\gcd(a_0, a_1, a_2, m) = 1$ and ζ is an m-th root of unity. This implies at once that

$$j(\mathbf{a}) \equiv 1 \pmod{(\zeta - 1)^3}.$$

Now suppose that $\gcd(a_0, a_1, a_2, m) > 1$. Then one can divide the a_i by a power of m_0 to get a character $\mathbf{a}' = [a_0', a_1', a_2']$ of degree m' such that $\gcd(a_0', a_1', a_2', m') = 1$, and the Ihara congruence says that

$$j_{m'}(\mathbf{a}') \equiv 1 \pmod{(\zeta_{m'}^{a_0'} - 1)(\zeta_{m'}^{a_1'} - 1)(\zeta_{m'}^{a_2'} - 1)},$$

where $\zeta_{m'}$ is an m'-th root of unity (hence a power of ζ). Recalling that $j(\mathbf{a}) = j_{m'}(\mathbf{a}')$, we get, once again, that

$$j(\mathbf{a}) \equiv 1 \pmod{(\zeta - 1)^3}.$$

To prove the congruence for higher-dimensional Jacobi sums, we exploit the inductive structure on $j(\mathbf{a})$ with respect to dimensions described in Lemma 1.3. As we noted above, if we "go down" the inductive structure we see that every higher-dimensional Jacobi sum can be expressed as a product of

- Jacobi sums of dimension 1,

- a power of q, and

- a power of $\chi(-1)$.

(Note that a Jacobi sum of dimension zero is simply equal to $\chi(-1)$.) Now, since m is odd, we have $\chi(-1) = 1$, and, since $q \equiv 1 \pmod{m}$, we know that $q \equiv 1 \pmod{(1 - \zeta)^{\phi(m)}}$. Since $m > 3$ we have $\phi(m) > 2$, and, together with Ihara's result for dimension 1, this gives the congruence we want.

Finally, writing $\mathfrak{J}(\mathbf{c}, \mathbf{a}) = \zeta^t j(\mathbf{a})$ with some t, $1 \leq t \leq m$, we get

$$\mathfrak{J}(\mathbf{c}, \mathbf{a}) - 1 = \zeta^t j(\mathbf{a}) - 1 = \zeta^t(j(\mathbf{a}) - 1) - (1 - \zeta^t) \equiv 0 \pmod{(1 - \zeta)}.$$

Notice that this computation in fact shows that

$$\mathfrak{J}(\mathbf{c}, \mathbf{a}) \equiv 1 \pmod{(1 - \zeta)^3}$$

also when m is not prime and $(t, m) > 1$. $\square$

Even for prime m, the precise nature of the $(1 - \zeta)$-adic expansion of Jacobi sums is still unknown. Still, partial results are available.

Suppose m is prime, and let $\pi = \zeta - 1$. Then (π) is a prime ideal in $L = \mathbb{Q}(\zeta)$ with $(\pi)^m = (m)$ and $\text{Norm}_{L/\mathbb{Q}}(-\pi) = m$. The (π)-adic completion of L is the local field $L_m := \mathbb{Q}_m(\zeta)$ equipped with a valuation ν_π (which extends the valuation ord_m of $\mathbb{Q}_m$ normalized by $\text{ord}_m(m) = 1$) such that $\nu_\pi(\pi) = 1$ and $\nu_\pi(m) = m - 1$. Let $\mathfrak{p}$ be a prime ideal in L over p such that $\text{Norm}_{L/\mathbb{Q}}(\mathfrak{p}) = q$. Then $q \equiv 1 \pmod{m}$ and $\mathbb{Z}[\zeta]/\mathfrak{p} \cong \mathbb{F}_q$. Let $\lambda : \mathbb{F}_q^\times / (\mathbb{F}_q^\times)^m \to \mathbb{Z}/m\mathbb{Z}$ be the isomorphism. Now we define for each i ($1 \le i \le (m-3)/2$),

$$\beta_{2i} := \sum_{x \in \mathbb{F}_q} \lambda(x)^{2i} \lambda(1-x) \in \mathbb{Z}/m\mathbb{Z}.$$

Further, we define functions Tlog and Texp by the truncated power series of the classical logarithm and exponential:

$$\text{Tlog}\, X = \sum_{i=1}^{m-1} (-1)^{i-1} \frac{(X-1)^{i-1}}{i} \in \mathbb{Z}_m[X]$$

and

$$\text{Texp}\, X = \sum_{i=0}^{m-1} \frac{X^i}{i!} \in \mathbb{Z}_m[X].$$

Let π' be an element of L defined by $\pi' \equiv \text{Tlog}\, \zeta \pmod{\pi^m}$. Then

$$\pi' \equiv \pi - \frac{\pi^2}{2} + \frac{\pi^3}{3} - \cdots + \frac{\pi^{m-1}}{m-1} \pmod{\pi^m} \equiv \pi u \pmod{\pi^m}$$

where u is a unit in L_m. Then π' is a prime element in L_m, so that $(\pi')^{m-1} = (m)$ and $\text{Norm}(\pi') = m \, \text{Norm}(u)$, with $\text{Norm}(u)$ relatively prime to m.

Now we make use of the m-adic expansion of Jacobi sums proved by Miki [Mik87] (cf. also [Yu94]). Let $\mathbf{a} = (a_1, a_2, \cdots, a_{n+1}) \in \mathfrak{A}_n^m$. Then

$$j(\mathbf{a}) \equiv \frac{1}{q} \text{Texp}\, Y \pmod{(\pi')^m}$$

where

$$Y = \sum_{i=1}^{(m-3)/2} [(\sum_{j=0}^{n+1} a_j^{2i+1}) \beta_{2i} \frac{(\pi')^{2i+1}}{(2i+1)!}] + \frac{q-1}{2m} (\sum_{j=0}^{n+1} a_j^{m-1})(\pi')^{m-1}$$

(where the a_i and the β_{2i} are lifted to $\mathbb{Z}$ in any way, since two such liftings are congruent mod $(\pi')^{m-1}$).

From this, we easily obtain:

THEOREM 1.8
Let $m > 3$ be a prime number, and fix a character $\mathbf{a} = (a_0, a_1, \cdots, a_{n+1}) \in \mathfrak{A}_n^m$. Suppose that there is an integer i such that $1 \le i \le (m-3)/2$ satisfying

$$(\sum_{j=0}^{n+1} a_j^{2i+1}) \beta_{2i} \not\equiv 0 \pmod{m}.$$

Let i_0 be the least such i. Then

$$\operatorname{ord}_\pi(j(\mathbf{a}) - 1) = 2i_0 + 1 \geq 3.$$

(For $i_0 = 1$, this is the Iwasawa congruence in [Iw75].) See also [Yu94].

Proof: From the above discussion, we have

$$\operatorname{ord}_{\pi'}(j(\mathbf{a}) - 1) = 2i_0 + 1 \geq 3,$$

and since $\pi' = \pi\, u$ with a unit u, the assertion follows. $\square$

One can also see from Miki's formula that the first few terms of the π'-adic expansion of $j(\mathbf{a}) - 1$ involve only odd powers of π'. On the other hand, the formula gives no information about further terms in the expansion. In particular, if we have $(\sum_{j=0}^{n+1} a_j^{2i+1})\beta_{2i} = 0$ for all i such that $1 \leq i \leq (m - 3)/2$, all we can conclude is that $\operatorname{ord}_\pi(j(\mathbf{a}) - 1) \geq m - 1$. That one can have equality here is shown by the example $m = 5$, $n = 8$, $p = 11$, $\mathbf{a} = (1,1,1,1,1,1,1,1,1,1)$—see Example 6.13 on page 71 below.

In spite of this, it would not be too far-off to expect that the following would be true:

CONJECTURE 1.9 *Fix m, n, and q as above, but assume $m > 3$ is prime and $n = 2d$ is even. Let $\mathbf{a} = (a_0, a_1, \cdots, a_{n+1}) \in \mathfrak{A}_n^m$. Then*

$$\operatorname{ord}_\pi(j(\mathbf{a}) - q^d) \quad \text{is always odd.}$$

The situation for twisted Jacobi sums is simpler, at least for prime degree m, since in this case the Iwasawa congruence clearly implies that

$$\operatorname{ord}_{(1-\zeta)}(\mathfrak{J}(\mathbf{c}, \mathbf{a}) - 1) = 1 \quad \text{unless} \quad \mathfrak{J}(\mathbf{c}, \mathbf{a}) = j(\mathbf{a})$$

(i.e., unless $\mathbf{c}^{\mathbf{a}}$ is an m-th power).

Finally, we note one useful consequence of the Iwasawa–Ihara congruence:

PROPOSITION 1.10
Let m be a prime power, $m > 3$, and let $n = 2d$ be even. Suppose $\mathfrak{J}(\mathbf{c}, \mathbf{a}) = q^d$. Then either

1. *$j(\mathbf{a}) = q^d$ and $\mathbf{c}^{\mathbf{a}} \in (k^\times)^m$, or*

2. *$\mathbf{c}^{\mathbf{a}} \notin (k^\times)^m$, but has order strictly less than m as an element of $k^\times/(k^\times)^m$.*

If m is prime, we have $\mathfrak{J}(\mathbf{c}, \mathbf{a}) = q^d$ if and only if $j(\mathbf{a}) = q^d$ and $\mathbf{c}^{\mathbf{a}} \in (k^\times)^m$.

Proof: Let $\chi(\mathbf{c}^{\mathbf{a}}) = \xi$. Then $\mathfrak{J}(\mathbf{c}, \mathbf{a}) = q^d$ if and only if $j(\mathbf{a}) = \xi q^d$. Since $q \equiv 1 \pmod{m}$, this implies that $j(\mathbf{a}) \equiv \xi \pmod{m}$. On the other hand, the Iwasawa–Ihara congruence says that $j(\mathbf{a}) \equiv 1 \pmod{(1 - \zeta)^2}$. It follows that $(1 - \xi)$ is divisible by $(1 - \zeta)^2$, which can only happen if ξ is not a primitive m-th root of unity, hence if $\mathbf{c}^{\mathbf{a}}$ has order strictly less than m. $\square$

REMARK 1.11 When m is not prime, one can indeed have $j(\mathbf{a}) \neq q^d$, but $\mathcal{J}(\mathbf{c}, \mathbf{a}) = q^d$. This occurs when $\chi(\mathbf{c}^{\mathbf{a}}) = j(\mathbf{a})/q^d$. We illustrate this phenomenon with an example. Take $m = 9$, $n = 6$, $p \equiv 1$ (mod 9), and choose

$$\mathbf{a} = [1, 3, 4, 4, 5, 6, 6, 7] \quad \text{and} \quad \mathbf{c} = [1, 2, 1, 1, 1, 1, 1, 1].$$

Then, setting $\zeta = e^{2\pi i/9}$, we have

$$j(\mathbf{a}) = \zeta^3 \, 19^3 \quad \text{and} \quad \mathcal{J}(\mathbf{c}, \mathbf{a}) = 19^3.$$

The importance of the case when $\mathcal{J}(\mathbf{c}, \mathbf{a}) = q^{n/2}$ should become clear as we go on to look at the zeta-function of $\mathcal{V}$. We begin by recalling the basic facts about the zeta-functions associated to our varieties $\mathcal{V} = \mathcal{V}_n^m(\mathbf{c})$. For each integer $i \geq 1$, let $k_i = \mathbb{F}_{q^i}$ and let N_i denote the number of k_i-rational points on $\mathcal{V} = \mathcal{V}_m^n(\mathbf{c})$. Then the zeta-function of $\mathcal{V} = \mathcal{V}_n^m(\mathbf{c})$ is defined as

$$Z(\mathcal{V}, T) = \exp\left(\sum_{i=1}^{\infty} \frac{N_i}{i} T^i\right) \in \mathbb{Q}((T)).$$

The following properties of $Z(\mathcal{V}, T)$ are well known:

1. The zeta-function $Z(\mathcal{V}, T)$ is a rational function of the form

$$Z(\mathcal{V}, T) = \frac{Q(\mathcal{V}, T)^{(-1)^{n+1}}}{\prod_{i=0}^{n}(1 - q^i T)} \in \mathbb{Q}(T)$$

   where $Q(\mathcal{V}, T) \in 1 + T\mathbb{Z}[T]$ with $\deg(Q) = \frac{m-1}{m}\{(m-1)^{n+1} + (-1)^{n+2}\}$.

2. Over $\mathbb{Q}(\zeta)$, $Q(\mathcal{V}, T)$ factors as

$$Q(\mathcal{V}, T) = \prod_{\mathbf{a} \in \mathfrak{A}_n^m} (1 - \mathcal{J}(\mathbf{c}, \mathbf{a})T)$$

   where

$$\mathcal{J}(\mathbf{c}, \mathbf{a}) = \bar{\chi}(c_0^{a_0} c_1^{a_1} \cdots c_{n+1}^{a_{n+1}}) j(\mathbf{a})$$

   is a twisted Jacobi sum of dimension n and of degree m with absolute value $q^{n/2}$.

The fact that the zeta-function has this form follows from the Davenport–Hasse relation on twisted Jacobi sums, which we recall briefly here. (See Davenport and Hasse [DH35], see also Weil [We49, We52].) For each k_i, define the characters χ_i and ψ_i by

$$\chi_i(z) = \chi(\text{Norm}_{k_i/k}(z)), \quad \psi_i(z) = \psi(\text{Trace}_{k_i/k}(z)).$$

Then the Gauss sum relative to k_i is

$$G(\chi_i) = G(\chi_i, \psi_i) := \sum_{x \in k_i} \psi_i(x) \chi_i(x).$$

The Davenport–Hasse relation describes the effect of base change on Gauss sums, that is,

$$G(\chi_i) = (-1)^{i-1}G(\chi).$$

From this and part 2 of Proposition 1.2, we can deduce a relation among twisted Jacobi sums under base change. Let $\mathcal{J}_i(\mathbf{c}, \mathbf{a})$ denote a twisted Jacobi sum relative to k_i and χ_i, ψ_i with $\mathcal{J}_1(\mathbf{c}, \mathbf{a}) = \mathcal{J}(\mathbf{c}, \mathbf{a})$. Then

$$\mathcal{J}_i(\mathbf{c}, \mathbf{a}) = \mathcal{J}(\mathbf{c}, \mathbf{a})^i \quad \text{for any } i \geq 1.$$

Furthermore,

$$N_i = 1 + q^i + \cdots + q^{in} + \sum \mathcal{J}_i(\mathbf{c}, \mathbf{a}).$$

In what follows, we will be interested in the special value of the zeta-function at $T = q^{-r}$ for various integers r. We set this up by writing

$$Q(\mathcal{V}, T) = (1 - q^r T)^\sigma \prod (1 - \mathcal{J}(\mathbf{c}, \mathbf{a})T),$$

where the product is now taken only over those $\mathbf{a}$ for which $\mathcal{J}(\mathbf{c}, \mathbf{a}) \neq q^r$ (note that the equality can only occur if n is even and $r = n/2$.) Then we have

$$Q(\mathcal{V}, q^{-s}) \sim (1 - q^{r-s})^\sigma \prod (1 - \mathcal{J}(\mathbf{c}, \mathbf{a})q^{-r})$$

as $s \to r$. It is this last product we are particularly interested in computing. It is useful to note that as $\mathbf{a}$ runs over the characters the twisted Jacobi sums will run over full Galois conjugacy classes in $\mathbb{Q}(\zeta)$, so that the product can be broken up as a product of norms; we will consider this fact more carefully in Chapter 6.

2 Cohomology groups of $\mathcal{V} = \mathcal{V}_n^m(\mathbf{c})$

The geometry and topology of $\mathcal{V} = \mathcal{V}_n^m(\mathbf{c})$ are closely linked to those of the Fermat variety $\mathcal{X} = \mathcal{V}_n^m(\mathbf{1})$, to which it is of course isomorphic over the algebraic closure $\bar{k}$. In fact, the phrase "geometric and topological invariants" of $\mathcal{V}$ usually refers to quantities depending only on the base-change of $\mathcal{V}$ to $\bar{k}$, which are therefore independent of the twisting vector $\mathbf{c} = (c_0, c_1, \ldots, c_{n+1})$ of the defining equation for $\mathcal{V}$. Some examples are various cohomological constructions. We record this for future reference.

LEMMA 2.1
Let $\mathcal{V} = \mathcal{V}_n^m(\mathbf{c})$ denote the diagonal hypersurface as above. The following cohomological constructions are independent (up to isomorphism) of the twisting vector $\mathbf{c} = (c_0, c_1, \ldots, c_{n+1})$:

1. for each prime $\ell \neq p = \mathrm{char}(k)$ and each $i \in \mathbb{Z}$, the ℓ-adic étale cohomology group $H^i(\mathcal{V}, \mathbb{Q}_\ell(i))$;

2. for each prime $\ell \neq p = \mathrm{char}(k)$ with $(\ell, m) = 1$ and each $i \in \mathbb{Z}$, the ℓ-adic étale cohomology group $H^i(\mathcal{V}, \mathbb{Z}_\ell(i))$;

3. for each pair (i, j), the de Rham cohomology group $H^{i+j}_{\mathrm{DR}}(\mathcal{V}/k)$, and the Hodge spectral sequence

$$E_1^{i,j} = H^j(\mathcal{V}, \Omega_\mathcal{V}^j) \Rightarrow H^{i+j}_{\mathrm{DR}}(\mathcal{V}/k);$$

4. for each $i \in \mathbb{Z}$, the crystalline cohomology groups

$$H^i(\mathcal{V}/W) \quad \text{and} \quad H^i(\mathcal{V}/W)_K;$$

5. for each pair (i, j), the Hodge–Witt cohomology groups $H^j(\mathcal{V}, W\Omega^i)$;

6. for each i, the formal groups $\Phi_\mathcal{V}^{i,n-i}$ arising from the Hodge–Witt cohomology groups $H^{n-i}(\mathcal{V}, W\Omega^i)$, and especially the Artin–Mazur formal group $\Phi_\mathcal{V}^\bullet = H^\bullet(\mathcal{V}, \hat{\mathbb{G}}_m)$.

Now we can use the known facts relating the cohomology of $\mathcal{V}$ to Jacobi sums to obtain some of the invariants of $\mathcal{V}$. We recall the definitions:

DEFINITION 2.2   *1. The i-th Betti number of $\mathcal{V}$, denoted $B_i(\mathcal{V})$, is defined by*

$$B_i(\mathcal{V}) = \begin{cases} \dim_{\mathbb{Q}_\ell} H^i(\mathcal{V}_{\bar{k}}, \mathbb{Q}_\ell(i)) & \text{if } \ell \neq p \\ \dim_K H^i(\mathcal{V}/W)_K & \text{if } \ell = p. \end{cases}$$

2. The (i, j)-th Hodge number of $\mathcal{V}$, denoted $h^{i,j}(\mathcal{V})$, is defined by

$$h^{i,j}(\mathcal{V}) = \dim_k H^j(\mathcal{V}, \Omega^i).$$

In particular, $h^{0,n}(\mathcal{V})$ is the geometric genus, $p_g(\mathcal{V})$, of $\mathcal{V}$.

The Hodge numbers of $\mathcal{V}$ are

$$h^0 = h^{0,n}(\mathcal{V}), \; h^1 = h^{1,n-1}(\mathcal{V}), \; \ldots, \; h^n = h^{n,0}(\mathcal{V}) \; .$$

The Hodge polygon of $\mathcal{V}$ is the polygon in $\mathbb{R}^2$ obtained by joining successively the line segments with slope i connecting the points

$$\left(\sum_{j=0}^{i-1} h^j, \sum_{j=0}^{i-1} j h^j \right) \quad and \quad \left(\sum_{j=0}^{i} h^j, \sum_{j=0}^{i} j h^j \right)$$

for each i, $0 \leq i \leq n$ (with the convention that the empty sum equals zero, so that the first point is the origin).

3. The slopes of $\mathcal{V}$ are defined to be the slopes of the isocrystal $H^n(\mathcal{V}/W)_K$. Let

$$\underbrace{\alpha_0, \ldots, \alpha_0}_{m_0}, \quad \underbrace{\alpha_1, \ldots, \alpha_1}_{m_1}, \quad \cdots, \quad \underbrace{\alpha_t, \ldots, \alpha_t}_{m_t}$$

be the slope sequence of $\mathcal{V}$, ordered so that

$$0 \leq \alpha_0 < \alpha_1 < \cdots < \alpha_t \leq n$$

and m_i denotes the multiplicity of slope α_i, respectively. Then the Newton polygon of $\mathcal{V}$ is the polygon in $\mathbb{R}^2$ obtained by joining successively the line segments with slope α_i connecting the points $(\sum_{j=0}^{i-1} m_j, \sum_{j=0}^{i-1} \alpha_j m_j)$ and $(\sum_{j=0}^{i} m_j, \sum_{j=0}^{i} \alpha_j m_j)$ for each i, $0 \leq i \leq t$ (with the same convention as to the empty sum).

The relation of the cohomology of $\mathcal{V}$ with twisted Jacobi sums allows us to reduce many questions regarding these invariants to combinatorial questions about the character vectors **a**. The following lemma does this for some of the cohomological invariants we have just defined.

LEMMA 2.3

1. *The i-th Betti number of $\mathcal{V}$ is computed by*

$$
B_i(\mathcal{V}) = \begin{cases} 0 & \text{if } i \text{ odd and } i \neq n \\ 1 & \text{if } i \text{ even and } i \neq n \\ 1 + \#\mathfrak{A}_n^m & \text{if } i = n \text{ even} \\ \#\mathfrak{A}_n^m & \text{if } i = n \text{ odd} \end{cases}
$$

and $\#\mathfrak{A}_m^n = \{(m-1)^{n+2} \pm (m-1)\}/m$, with the sign depending on whether n is even or odd.

2. *The (i,j)-th Hodge number of $\mathcal{V}$ is computed by*

$$
h^{i,j}(\mathcal{V}) = \begin{cases} 0 & \text{if } i+j \neq n \text{ and } i \neq j \\ 1 & \text{if } i+j \neq n \text{ and } i = j \\ 1 + \#\{\mathbf{a} \in \mathfrak{A}_n^m \mid \|\mathbf{a}\| = i\} & \text{if } i+j = n \text{ and } n \text{ even} \\ \#\{\mathbf{a} \in \mathfrak{A}_n^m \mid \|\mathbf{a}\| = i\} & \text{if } i+j = n \text{ and } n \text{ odd.} \end{cases}
$$

Furthermore, we have $B_n(\mathcal{V}) = \sum_{i=0}^n h^{i,n-i}(\mathcal{V})$.

3. *Let $\mathfrak{p}$ be a prime in L lying above p, let H be a decomposition subgroup for $\mathfrak{p}$, and let $\mathrm{Norm}(\mathfrak{p}) = p^f$. Then the slopes of $\mathcal{V}$ are the numbers*

$$
\{A_H(\mathbf{a})/f \mid \mathbf{a} \in \mathfrak{A}_n^m\}
$$

arranged in increasing manner.

Proof: (Cf. Koblitz [Ko75], Suwa and Yui [SY88], and Suwa [Su91a, Su91b].) Given what is known about the Fermat variety $\mathcal{X}$ (see Suwa and Yui [SY88], Suwa [Su91a, Su91b]), we have only to explain the assertion on the slopes of $\mathcal{V}$. The eigenvalues of the Frobenius endomorphism of $\mathcal{V}$ differ from those of $\mathcal{X}$ just by the m-th roots of unity

$$
\bar{\chi}(c_0^{a_0} \cdots c_{n+1}^{a_{n+1}}) \quad \text{with} \quad \mathbf{a} \in \mathfrak{A}_n^m.
$$

Therefore, the p-adic ordinals of the eigenvalues of $\mathcal{V}$ are the same as those for the Fermat variety $\mathcal{X}$. □

We will later compute explicitly these invariants in a few specific cases. We also recall, from [Ma72],

THEOREM 2.4 (MAZUR)
The Newton polygon of $\mathcal{V}$ lies above or on the Hodge polygon of $\mathcal{V}$.

We now consider formal groups arising from $\mathcal{V}$, e.g., the Artin–Mazur formal groups $\Phi_{\mathcal{V}}^\bullet = H^\bullet(\mathcal{V}, \widehat{\mathbb{G}}_m)$ of $\mathcal{V}$.

LEMMA 2.5

There is a connected smooth formal group $\Phi_{\mathcal{V}}^{i,n-i}$ over k whose Cartier module is isomorphic to $H^{n-i}(\mathcal{V}, W\Omega^i)$. In particular, the Artin–Mazur functor $\Phi_{\mathcal{V}}^n = H^n(\mathcal{V}, \widehat{\mathbb{G}}_m)$ is representable by a connected smooth formal group $\Phi_{\mathcal{V}}^{0,n}$ over k of dimension $p_g(\mathcal{V})$. Furthermore, $\Phi_{\mathcal{V}}^{i,n-i}$ has the following properties:

1. $\Phi_{\mathcal{V}}^{i,n-i}$ is isomorphic over $\bar{k}$ to the corresponding formal group of the Fermat variety, $\Phi_{\mathcal{X}}^{i,n-i}$.

2. There is a canonical exact sequence of connected smooth formal groups

$$0 \longrightarrow \mathcal{U}_{\mathcal{V}}^{i,n-i} \longrightarrow \Phi_{\mathcal{V}}^{i,n-i} \longrightarrow \mathcal{D}_{\mathcal{V}}^{i,n-i} \longrightarrow 0$$

where $\mathcal{U}_{\mathcal{V}}^{i,n-i}$ is unipotent and $\mathcal{D}_{\mathcal{V}}^{i,n-i}$ is p-divisible, whose dimension and the height are explicitly given as follows:

$$\dim \mathcal{D}_{\mathcal{V}}^{i,n-i} = \sum_{\substack{\mathbf{a} \in \mathfrak{A}_m^n \\ i \leq A_H(\mathbf{a})/f < i+1}} ((i+1) - A_H(\mathbf{a})/f))$$

$$\dim \mathcal{U}_{\mathcal{V}}^{i,n-i} = T^{i,n-i}(\mathcal{V}) = \dim H^{n-i}(\mathcal{V}, W\Omega^i)/V,$$

and

$$\mathrm{ht}\, \mathcal{D}_{\mathcal{V}}^{i,n-i} = \#\{\mathbf{a} \in \mathfrak{A}_m^n \mid i \leq A_H(\mathbf{a})/f < i+1\}.$$

Proof: (See Artin and Mazur [AM77], and also Suwa and Yui [SY88].) Since the Hodge–Witt cohomology groups $H^{n-i}(\mathcal{V}, W\Omega^i)$ of the twisted diagonal hypersurface are isomorphic to the Hodge–Witt cohomology groups $H^{n-i}(\mathcal{X}, W\Omega^i)$ of the Fermat hypersurface, the assertion follows from the work of Suwa and Yui [SY88, Chapter 3]. (See also Suwa [Su91a, Su91b].) $\square$

3 Twisted Fermat motives

Let $\mathcal{V} = \mathcal{V}_n^m(\mathbf{c})$ be a diagonal hypersurface with twist $\mathbf{c}$ over $k = \mathbb{F}_q$. The polynomial $Q(\mathcal{V}, T)$ has degree $\#\mathfrak{A}_n^m$ (essentially the n-th Betti number), which is in general a very large number. However, one sees easily that it factors very highly over $\mathbb{Z}$. This allows us to break up the problem of studying special values into a series of similar problems for factors of $Q(\mathcal{V}, T)$. Formally, this is done by introducing twisted Fermat motives, which turn out to be attached to certain quotients of $\mathcal{V}$. Here we regard the group $\mathfrak{G} = \mathfrak{G}_n^m$ as a subgroup of the automorphism group $\mathrm{Aut}(\mathcal{V})$ of $\mathcal{V}$.

DEFINITION 3.1 *For any* $\mathbf{a} \in \widehat{\mathfrak{G}}$, *let*

$$p_{\mathbf{a}} = \frac{1}{\#\mathfrak{G}} \sum_{g \in \mathfrak{G}} \mathbf{a}(g)^{-1} g = \frac{1}{m^{n+1}} \sum_{g \in \mathfrak{G}} \mathbf{a}(g)^{-1} g \in \mathbb{Z}[\frac{1}{m}, \zeta][\mathfrak{G}].$$

Recall that $(\mathbb{Z}/m\mathbb{Z})^\times$ acts on $\widehat{\mathfrak{G}}$ by

$$t \cdot (a_0, a_1, \ldots, a_{n+1}) = (ta_0, ta_1, \ldots, ta_{n+1}),$$

and that this action is related to the Galois action on the twisted Jacobi sum corresponding to $\mathbf{a} = (a_0, a_1, \ldots, a_{n+1})$. This suggests we consider the $(\mathbb{Z}/m\mathbb{Z})^\times$-orbit of $\mathbf{a}$, denoted $A = [\mathbf{a}]$. (It will be relevant later to note that the order of A is at most $\phi(m)$.) Let

$$p_A = \sum_{\mathbf{a} \in A} p_{\mathbf{a}} \in \mathbb{Z}[1/m][\mathfrak{G}].$$

Then it is easily seen that $p_{\mathbf{a}}$ and p_A are idempotents, and that

$$\sum_{\mathbf{a} \in \widehat{\mathfrak{G}}} p_{\mathbf{a}} = \sum_{A \in O(\widehat{\mathfrak{G}})} p_A = 1,$$

where $O(\widehat{\mathfrak{G}})$ denotes the set of $(\mathbb{Z}/m\mathbb{Z})^\times$-orbits in $\widehat{\mathfrak{G}}$. Identifying an automorphism $g \in \mathfrak{G} \subset \mathrm{Aut}(\mathcal{V})$ with its graph Γ_g, we see that $p_A \in \mathrm{End}(\widehat{\mathcal{V}}) \otimes \mathbb{Z}[1/m]$ may be regarded as an algebraic cycle on $(\mathcal{V} \times \mathcal{V})_k$ with coefficients in $\mathbb{Z}[1/m]$. Therefore, the pair $(\mathcal{V}, p_A) := \mathcal{V}_A$ defines a motive over k, corresponding to the $(\mathbb{Z}/m\mathbb{Z})^\times$-orbit of A in $\widehat{\mathfrak{G}}$.

The same projector p_A defines the Fermat motive $\mathcal{M}_A$ of dimension n and of degree m corresponding to $A = [\mathbf{a}]$ (Shioda [Sh87]). Thus $\mathcal{V}_A$ is a twisted version of the Fermat motive $\mathcal{M}_A$. When $\mathbf{c}$ is fixed, we call $\mathcal{V}_A$ the *twisted Fermat motive of dimension n and of degree m corresponding to $A = [\mathbf{a}]$.* (For a general background on motives, see for instance Soulé [So84].)

This construction gives a decomposition of the motive attached to the variety $\mathcal{V}$, as follows:

$$\tilde{\mathcal{V}} = (\mathcal{V}, \Delta_\mathcal{V}) = \bigoplus_{A \in O(\widehat{\mathfrak{G}})} (\mathcal{V}, p_A) = \bigoplus_{A \in O(\widehat{\mathfrak{G}})} \mathcal{V}_A$$

corresponding to $\sum p_A = 1$. We call this the *motivic decomposition* of $\mathcal{V}$. In cohomology, this corresponds to the decomposition

$$H^n(\mathcal{V}) = \bigoplus H^n(\mathcal{V})(A) = \bigoplus H^n(\mathcal{V}_A),$$

where H^n denotes any of the cohomology theories mentioned above, and where $H^n(\mathcal{V})(A)$ is the part of the cohomology group fixed by the kernel of $\mathbf{a}$. If we decompose $H^n(\mathcal{V}) \otimes L$ according to the characters of $\mathfrak{G}$, so that

$$H^n(\mathcal{V}) \otimes L = \bigoplus_{\mathbf{a} \in \widehat{\mathfrak{G}}} H^n(\mathcal{V})(\mathbf{a}),$$

then we have

$$H^n(\mathcal{V})(A) = H^n(\mathcal{V}) \bigcap \bigoplus_{\mathbf{a} \in A} H^n(\mathcal{V})(\mathbf{a}).$$

It is interesting to relate the motive $\mathcal{V}_A$ to a "real" geometric object. (Cf. Schoen [So90].) This is not hard to do, since it suffices to construct the quotient $\mathcal{V}$ by an appropriate subgroup of $\mathfrak{G}$. Let $X_0, X_1, \ldots, X_{n+1}$ be homogeneous coordinates on $\mathbb{P}_k^{n+1}$ and consider the hyperplane $\mathcal{H}$ defined by

$$c_0 X_0 + c_1 X_1 + \cdots + c_{n+1} X_{n+1} = 0. \tag{$*$}$$

Then the morphism

$$\mathbb{P}_k^{n+1} \to \mathbb{P}_k^{n+1}$$
$$(X_0, X_1, \ldots, X_{n+1}) \mapsto (X_0^m, X_1^m, \ldots, X_{n+1}^m)$$

realizes $\mathcal{V} = \mathcal{V}_n^m$ as a finite Galois cover of $\mathcal{H}$ with Galois group $\mathfrak{G}$. The branch locus consists of the $(n+2)$-hyperplanes $X_i = 0$ for $i = 0, 1, \ldots, n+1$. Now for each character $\mathbf{a} \in \widehat{\mathfrak{G}}$, let $\mathfrak{G}_\mathbf{a}$ denote the kernel of the map $\mathbf{a} : \mathfrak{G} \to \mu_m$ given by $g \mapsto \mathbf{a}(g)$, i.e.,

$$\mathfrak{G}_\mathbf{a} = \{ g \in \mathfrak{G} \mid \mathbf{a}(g) = 1 \}.$$

(Note that this depends only on the $(\mathbb{Z}/m\mathbb{Z})^\times$-orbit of $\mathbf{a}$.) Then of course $\mathfrak{G}/\mathfrak{G}_\mathbf{a} = \mathrm{Im}(\mathbf{a}) \subset (\mathbb{Z}/m\mathbb{Z})$.

PROPOSITION 3.2
The quotient $\mathfrak{G}_{\mathbf{a}} \backslash \mathcal{V}$ is the normalization of the complete intersection in $\mathbb{P}^{n+2}$ given by the equations

$$Y^m = \prod_{i=0}^{n+1} X_i^{a_i}, \quad \sum_{i=0}^{n+1} c_i X_i = 0. \tag{$\dagger$}$$

Proof: This is essentially clear. Let W_0 denote the complete intersection above. Then $\mathcal{V}$ maps to W_0 via

$$(X_0, X_1, \ldots, X_{n+1}) \mapsto (X_0^m, X_1^m, \ldots, X_{n+1}^m, \prod X_i^{a_i})$$

and W_0 maps to the hyperplane $\mathcal{H}$ by projection on the first $n+2$ coordinates. It is a trivial matter to see that $\mathfrak{G}_{\mathbf{a}}$ acts trivially on W_0, and that no other elements of $\mathfrak{G}$ do. The rest follows. $\square$

Thus, if $\mathcal{W}_A$ denotes the quotient, we have

$$H^n(\mathcal{V})(A) \cong H^n(\mathcal{W}_A)$$

for each of the cohomology theories above and each $n \geq 1$. It might be interesting to explore these varieties further.

LEMMA 3.3
The Frobenius endomorphism Φ of $\mathcal{V}$ relative to k commutes with the motivic decomposition. That is, the endomorphism Φ^* induced from the Frobenius endomorphism on the cohomology groups defined in Lemma 2.1 acts semisimply.

Proof: Note that
$$\Phi^* \cdot p_A = p_A \cdot \Phi^*. \quad \square$$

LEMMA 3.4
The polynomial $Q(\mathcal{V}, T)$ factors as

$$Q(\mathcal{V}, T) = \prod_{A \in O(\mathfrak{G})} Q(\mathcal{V}_A, T)$$

where

$$Q(\mathcal{V}_A, T) := \prod_{\mathbf{a} \in A} (1 - \mathfrak{J}(\mathbf{c}, \mathbf{a})T) \in 1 + \mathbb{Z}[T]$$

is the polynomial, not necessarily irreducible over $\mathbb{Q}$, corresponding to the twisted Fermat motive $\mathcal{V}_A$.

The numerical and geometric invariants of $\mathcal{V}_A$ are defined in the obvious way, and their values can be computed analogously to those of $\mathcal{V}$, in terms of twisted Jacobi sums.

LEMMA 3.5

1. *The i-th Betti number of $\mathcal{V}_A$ is*

$$
B_i(\mathcal{V}_A) = \dim_{\mathbb{Q}_\ell} H^i(\mathcal{V}_{A_{\bar{k}}}, \mathbb{Q}_\ell) = \dim_K H^i(\mathcal{V}_A/W)_K
$$
$$
= \begin{cases}
\#A & \text{if } i = n \text{ and } A \subset \mathfrak{A}_n^m \\
1 & \text{if } i \text{ even and } A = [(0, \ldots, 0)] \\
0 & \text{otherwise.}
\end{cases}
$$

We have $B_n(\mathcal{V}_A) \le \phi(m)$, and the equality holds whenever $\gcd(\mathbf{a}, m) = \gcd((a_0, a_1, \ldots, a_{n+1}), m) = 1$ (hence in particular when m is prime). Moreover, we have

$$
B_n(\mathcal{V}) = \sum_{A \in O(\widehat{\mathfrak{G}})} B_n(\mathcal{V}_A).
$$

2. *The (i, j)-th Hodge number of $\mathcal{V}_A$ is*

$$
h^{i,j}(\mathcal{V}_A) := \dim_k H^j(\mathcal{V}_A, \Omega^i)
$$
$$
= \begin{cases}
\#\{\, \mathbf{a} \in A \mid \|\mathbf{a}\| = i \,\} & \text{if } i + j = n \text{ and } A \subset \mathfrak{A}_n^m \\
1 & \text{if } A = [(0, \ldots, 0)] \text{ and } i = j \\
0 & \text{otherwise}
\end{cases}
$$

and moreover, we have

$$
h^{i,j}(\mathcal{V}) = \sum_{A \in O(\widehat{\mathfrak{G}})} h^{i,j}(\mathcal{V}_A) .
$$

The Hodge numbers of $\mathcal{V}_A$ are defined by

$$
h^0(\mathcal{V}_A) = h^{0,n}(\mathcal{V}_A),\ h^1(\mathcal{V}_A) = h^{1,n-1}(\mathcal{V}_A),\ \ldots,\ h^n(\mathcal{V}_A) = h^{n,0}(\mathcal{V}_A) .
$$

In particular, $h^{0,n}(\mathcal{V}_A)$ is the geometric genus, $p_g(\mathcal{V}_A)$, of $\mathcal{V}_A$. Furthermore, we have $\sum_{i=0}^{n} h^{i,n-i}(\mathcal{V}_A) = B_n(\mathcal{V}_A)$.

3. *The slopes of $\mathcal{V}_A$ are the slopes of the isocrystal $H^n(\mathcal{V}_A/W)_K$, and are given by*

$$
\{\, A_H(\mathbf{a})/f \mid \mathbf{a} \in A \,\}
$$

arranged in increasing order.

4. *Mazur's theorem can be syphoned to motives, and indeed, the Newton polygon of $\mathcal{V}_A$ lies above or on the Hodge polygon of $\mathcal{V}_A$.*

5. The formal group $\Phi_{\mathcal{V}_A}^{i,n-i}$ of $\mathcal{V}_A$ is defined by the formal group whose Cartier module is isomorphic over k to the Hodge–Witt cohomology group $H^{n-i}(\mathcal{V}_A, W\Omega^i)$ for each i, $0 \leq i \leq n$. Let $\mathcal{D}_{\mathcal{V}_A}^{i,n-i}$ be the p-divisible part of $\Phi_{\mathcal{V}_A}^{i,n-i}$. Then

$$\dim \mathcal{D}_{\mathcal{V}_A}^{i,n-i} = \sum_{\substack{\mathbf{a} \in A \\ i \leq A_H(\mathbf{a})/f < i+1}} [(i+1) - A_H(\mathbf{a})/f],$$

$$\operatorname{codim} \mathcal{D}_{\mathcal{V}_A}^{i,n-i} = \sum_{\substack{\mathbf{a} \in A \\ i \leq A_H(\mathbf{a})/f < i+1}} (A_H(\mathbf{a})/f - i),$$

and

$$\operatorname{ht} \mathcal{D}_{\mathcal{V}_A}^{i,n-i} = \#\{\mathbf{a} \in A \mid i \leq A_H(\mathbf{a})/f < i+1\}.$$

Proof: The assertions in Lemmas 2.1, 2.3 and 2.5 are passed onto motives by Lemma 3.3. $\square$

We can make the following definitions:

DEFINITION 3.6

1. $\mathcal{V}_A$ is ordinary *if the Newton polygon coincides with the Hodge polygon of* $\mathcal{V}_A$.

2. $\mathcal{V}_A$ is supersingular *if the Newton polygon has the pure slope* $n/2$.

3. $\mathcal{V}_A$ is strongly supersingular *if* $\mathcal{J}(\mathbf{c}, \mathbf{a}) = q^{n/2}$ *for every* $\mathbf{a} \in A$.

4. $\mathcal{V}_A$ is of Hodge–Witt type *if* $H^j(\mathcal{V}_A, W\Omega^i)$ *is of finite type over* W *for any pair* (i, j) *with* $i + j = n$.

We will say a character $\mathbf{a}$ is ordinary (resp. Hodge–Witt, resp. supersingular) if the motive associated to its $(\mathbb{Z}/m\mathbb{Z})^\times$-orbit is ordinary (resp. Hodge–Witt, resp. supersingular).

LEMMA 3.7

If $\mathcal{V}_A$ is supersingular then $j(\mathbf{a}) = \xi q^{n/2}$, where ξ is an m-th root of unity. If m is a prime, $m > 3$, then in fact $j(\mathbf{a}) = q^{n/2}$.

Proof: The first assertion is well known. The second follows from Proposition 1.10. $\square$

From the lemma we see that if $\mathcal{V}_A$ is supersingular then $\mathcal{J}(\mathbf{c}, \mathbf{a})$ differs from $q^{n/2}$ by a factor of a root of unity. This explains the term "strongly supersingular" above.

The case of $p \equiv 1 \pmod{m}$, so that $f = 1$, is special in many ways. First of all, we have $A_H(\mathbf{a}) = \|\mathbf{a}\|$ for every character $\mathbf{a}$, and this immediately shows that in this case every motive will be ordinary. Moreover, in this case the fact that a character **a** is supersingular does not depend on the specific prime p (as it does in the general case). Finally, if a character **a** is supersingular with respect to any prime $p \equiv 1 \pmod{m}$, then it is clearly supersingular with respect to any other prime.

It will be useful for the subsequent discussions to give a combinatorial characterization of ordinary, resp., of Hodge–Witt type, resp., supersingular twisted Fermat motives.

PROPOSITION 3.8
Let $\mathcal{V}_A$ denote a twisted Fermat motive of dimension n and degree m.

 1. *The following conditions are equivalent.*

    (a) *$\mathcal{V}_A$ is ordinary.*
    (b) *$\|p\mathbf{a}\| = \|\mathbf{a}\|$ for any $\mathbf{a} \in A$.*

 2. *The following conditions are equivalent.*

    (a) *$\mathcal{V}_A$ is supersingular.*
    (b) *$A_H(\mathbf{a}) = nf/2$ for any $\mathbf{a} \in A$.*

 3. *The following conditions are equivalent.*

    (a) *$\mathcal{V}_A$ is of Hodge–Witt type.*
    (b) *$\|p^j\mathbf{a}\| - \|\mathbf{a}\| = 0, \pm 1$ for any $\mathbf{a} \in A$ and for any j, $0 < j < f$.*

Proof: (Cf. Suwa and Yui [SY88], Suwa [Su91a].) The assertions of (1) and (2) follow immediately from the definition. For (3), see Suwa and Yui [SY88], Chapter 3. □

REMARK 3.9

 1. One sees from the proposition that if $\mathcal{V}_A$ is ordinary, then $\mathcal{V}_A$ is automatically of Hodge–Witt type. However, the converse is not true. (See Illusie and Raynaud [IR83].)

 2. $\mathcal{V}_A$ can be ordinary or of Hodge–Witt type, and at the same time supersingular.

 3. The relations among these properties for Fermat motives M_A and for twisted Fermat motives $\mathcal{V}_A$ are as expected: if M_A is ordinary (resp. of Hodge–Witt type, resp. supersingular), then so is $\mathcal{V}_A$. This is, of course, clear from Proposition 3.8.

PROPOSITION 3.10
Let $\mathcal{V}_A$ be a twisted Fermat motive of degree m and of even dimension $n = 2d$. Then the following statements are equivalent:

1. $\mathcal{V}_A$ is ordinary and supersingular.

2. $\|\mathbf{a}\| = d$ for every $\mathbf{a} \in A$.

3. $h^{d,d}(\mathcal{V}_A) = B_n(\mathcal{V}_A)$.

Proof: Clear. □

The next result gives some information about motives that are ordinary and are not supersingular.

PROPOSITION 3.11
Let $\mathcal{V}_A$ be a twisted Fermat motive of degree m and of even dimension $n = 2d$. As above, let f denote the order of the decomposition group $H \subset G$ of an ideal dividing p. If $\mathcal{V}_A$ is ordinary but not supersingular, then

$$h^{d,d}(\mathcal{V}_A) \leq B_n(\mathcal{V}_A) - 2f$$

and there does not exist an integer j such that $p^j \equiv -1 \pmod{m}$.

Proof: Let H be the cyclic subgroup of $(\mathbb{Z}/m\mathbb{Z})^\times$ generated by p. Note, first that if $-1 \in H$ then every character $\mathbf{a}$ is supersingular, since the formula for $A_H(\mathbf{a})$ will break up into a sum of $f/2$ terms of the form $\|\mathbf{a}\| + \| - \mathbf{a}\|$, each of which equals $2d$. In particular, if $f = B_n(\mathcal{V}_A)$ then $\mathcal{V}_A$ is automatically supersingular, so that our statement does make sense.

Next, since $\mathcal{V}_A$ is ordinary, we have

$$\|\mathbf{a}\| = \|p\mathbf{a}\| = \cdots = \|p^{f-1}\mathbf{a}\| \quad \text{for every} \quad \mathbf{a} \in A.$$

On the other hand, $\mathcal{V}_A$ is not supersingular, so that

$$A_H(\mathbf{a}) = \|\mathbf{a}\| + \|p\mathbf{a}\| + \cdots + \|p^{f-1}\mathbf{a}\| \neq df \quad \text{for some} \quad \mathbf{a} \in A.$$

This implies that

$$\|\mathbf{a}\| \neq d \quad \text{for some} \quad \mathbf{a} \in A,$$

and therefore that $\|t\mathbf{a}\| \neq d$ for every $t \in H$. Now, since $\|\mathbf{a}\| + \| - \mathbf{a}\| = n = 2d$, we also have $\| - \mathbf{a}\| \neq d$, which again implies that $\| - t\mathbf{a}\| \neq d$ for all $t \in H$. This shows that there can be at most $B_n(\mathcal{V}_A) - 2f$ vectors $\mathbf{a}$ for which $\|\mathbf{a}\| = d$, which proves our claim. □

Note that the proof also shows that if -1 is a power of p modulo m, then every motive is supersingular. In particular, this will be the case when m is an odd prime power and f is even. Shioda and Katsura [SK79, Theorem 3.4] have shown that the converse is also true: if every motive is supersingular, then -1 is a power of p modulo m.

Characterizations of ordinary twisted Fermat motives and twisted Fermat motives of Hodge–Witt type in terms of formal groups can be deduced from Ekedahl's results in [Ek84].

PROPOSITION 3.12

Let $\mathcal{V}_A$ be a twisted Fermat motive of degree m and dimension n.

1. The following conditions are equivalent.

   (a) $\mathcal{V}_A$ is ordinary.

   (b) $\Phi_{\mathcal{V}_A}^{i,n-i}$ is isomorphic over $\bar{k}$ to the multiplicative group $\hat{\mathbb{G}}_{m,\bar{k}}$ for each i, $0 \leq i \leq (n-1)/2$.

2. The following conditions are equivalent.

   (a) $\mathcal{V}_A$ is of Hodge–Witt type.

   (b) $\Phi_{\mathcal{V}_A}^{i,n-i}$ is isomorphic over $\bar{k}$ to a p-divisible formal group for each i, $0 \leq i \leq (n-1)/2$.

   (c) $h^{i,n-i}(\mathcal{V}_A) = \dim \mathcal{D}^{i,n-i}\mathcal{V}_A + \operatorname{codim} \mathcal{D}_{\mathcal{V}_A}^{i-1,n-i+1}$ for each i, $0 \leq i \leq (n-1)/2$.

3. If $\mathcal{V}_A$ is supersingular, then $\Phi_{\mathcal{V}_A}^{i,n-i}$ is unipotent for every i.

We have computed the invariants of various twisted Fermat motives. A few examples of such computations can be found in Table A.2.

Passing to the global situation, we can now make the following definitions for diagonal hypersurfaces.

DEFINITION 3.13 *Let $\mathcal{V}$ be a diagonal hypersurface of dimension n and of degree m.*

1. $\mathcal{V}$ *is said to be* ordinary *if each twisted Fermat motive $\mathcal{V}_A$ is ordinary.*

2. $\mathcal{V}$ *is said to be* supersingular *if each twisted Fermat motive $\mathcal{V}_A$ is supersingular.*

3. $\mathcal{V}$ *is said to be* strongly supersingular *if each twisted Fermat motive $\mathcal{V}_A$ is strongly supersingular.*

4. $\mathcal{V}$ *is said to be of* Hodge–Witt type *if each twisted Fermat motive $\mathcal{V}_A$ is of Hodge–Witt type.*

REMARK 3.14 Most diagonal hypersurfaces are of mixed type. One easy case, however, was noted above: diagonal hypersurfaces of degree m are ordinary when $p \equiv 1 \pmod{m}$. Similarly, the theorem of Shioda and Katsura mentioned above says that diagonal hypersurfaces of degree m are supersingular when there exists an integer j such that $p^j \equiv -1 \pmod{m}$.

REMARK 3.15 As noted above, ordinary diagonal hypersurfaces are of Hodge–Witt type. While there are many examples of non-ordinary twisted Fermat motives of Hodge–Witt type, it turns out that there are very few examples of non-ordinary diagonal hypersurfaces of Hodge–Witt type. Indeed, Toki and Sasaki in [TS] have recently used the combinatorial characterization above to determine all non-ordinary diagonal hypersurfaces of Hodge–Witt type. They are those with

$$
\begin{aligned}
(m,n) &= (3,2) &&\text{and } p \equiv 2 \pmod 3 \\
&= (3,3) &&\text{and } p \equiv 2 \pmod 3 \\
&= (4,3) &&\text{and } p \equiv 3 \pmod 4 \\
&= (3,5) &&\text{and } p \equiv 2 \pmod 3 \\
&= (7,2) &&\text{and } p \equiv 2 \text{ or } 4 \pmod 3.
\end{aligned}
$$

REMARK 3.16 To simplify the calculations, one notes that many of our motives are isomorphic, reducing greatly the number of cases to be considered. First of all, note that two Fermat motives $\mathcal{M}_A$ and $\mathcal{M}_{A'}$ will be isomorphic whenever some character $\mathbf{a} = (a_0, a_1, \ldots, a_{n+1}) \in A$ is equal to a permutation of a character in A'. Thus, for computations which depend only on $\mathcal{M}_A$, one can simply work with a representative from each isomorphism class, and keep a count of their multiplicity.

Computations involving the twist $\mathbf{c}$ require a bit more care. If $\mathbf{c}$ is particularly simple however, similar ideas still apply. For example, if we have $\mathbf{c} = (c_0, 1, 1, \ldots, 1)$ with $c_0 \neq 1$, which is a case we will often want to consider, we need only break up the isomorphism classes according to the first entries in the characters, so that the motive $\mathcal{V}_A$ determined by $\mathbf{a} = (1,1,2,2,5,5)$ is isomorphic to that determined by $\mathbf{a} = (1,2,1,2,5,5)$, though not to that determined by $\mathbf{a} = (2,1,1,2,5,5)$. This only slightly complicates keeping track of the multiplicities.

4 The inductive structure and the Hodge and Newton polygons

Shioda [Sh79a, Sh82b] (cf. also Shioda and Katsura [SK79]) has studied geometrically the inductive structure of Fermat varieties. We have described the inductive structure of diagonal hypersurfaces in Lemma 1.3, Lemma 1.5, and Proposition 1.6. In this section, we shall consider cohomological realizations of the inductive structure by Hodge cohomology, étale (or crystalline) cohomology and Hodge–Witt cohomology. More concretely, we shall see the effect of the inductive structure on the Hodge polygon, the Newton polygon and the formal groups attached to motives $\mathcal{M}_A$ and $\mathcal{V}_A$.

The results in this chapter are well known, even in a more general setting, but it seemed interesting to record both the results and their (easy) proofs for the reader's convenience.

To begin, we review the inductive structure described in Chapter 1. Recall that $\mathfrak{A}_n^m$ denotes the set of character vectors $\mathbf{a} = (a_0, a_1, \ldots, a_{n+1})$ such that $\sum a_i \equiv 0 \pmod{m}$ and $a_i \not\equiv 0 \pmod{m}$ for each i. In particular, we have $\mathfrak{A}_0^m = \{(a, m - a) \mid a \in \mathbb{Z}/m\mathbb{Z}, a \not\equiv 0\}$.

Then the inductive structure described in Chapter 1 (specialized to the case when $r = n - 1$ and $s = 1$) gives a map

$$\mathfrak{A}_n^m \times \mathfrak{A}_0^m \mapsto \mathfrak{A}_{n+2}^m$$

by concatenation of vectors:

$$((a_0, a_1, \ldots, a_{n+1}), (a, m - a)) \mapsto (a_0, a_1, \ldots, a_{n+1}, a, m - a).$$

We call this inductive structure the *type I* inductive structure. We will refer to vectors in $\mathfrak{A}_{n+2}^m$ which are in the image of this map as *induced characters of type I*, and to twisted Fermat motives corresponding to such vectors as *type I motives*. We note that for each $\mathbf{a} \in \mathfrak{A}_n^m$, there are exactly $m - 1$ induced characters $\tilde{\mathbf{a}} \in \mathfrak{A}_{n+2}^m$ of type I. We view the inductive structure of type I as a sort of "tree" beginning at dimensions $n = 0$ and $n = 2$ and branching up through dimensions of the same parity (since each step adds two to the dimension).

39

These do not exhaust $\mathfrak{A}_{n+2}^m$, as Shioda [Sh79a] has shown. The complement, however, is also obtained from lower dimensions, specifically, dimensions $n + 1$ and 1; it is isomorphic to the subset $\mathfrak{A}_{n+1,1}^m$ of $\mathfrak{A}_{n+1}^m \times \mathfrak{A}_1^m$ defined by

$$\mathfrak{A}_{n+1,1}^m = \{((a_0, a_1, \ldots, a_{n+2}), (b_0, b_1, b_2)) \mid a_{n+2} + b_2 = m\}.$$

Then the inductive structure gives a map

$$\mathfrak{A}_{n+1,1}^m \mapsto \mathfrak{A}_{n+2}^m$$

by assigning to a pair $(\mathbf{a}, \mathbf{b})$ the vector $\mathbf{a}\#\mathbf{b} = (a_0, a_1, \ldots, a_{n+1}, b_0, b_1)$. We call this inductive structure the *type II* inductive structure. We will refer to vectors in $\mathfrak{A}_{n+2}^m$ which are in the image of this map as *induced characters of type II*, and twisted Fermat motives corresponding to such vectors as *type II motives*. We note that for each $\mathbf{a} \in \mathfrak{A}_{n+1}^m$, there are at most $m - 2$ induced characters of type II with $b_0 + b_1 = a_{n+2}$, and at most $m - a_{n+2} - 1$ induced characters of type II with $b_0 + b_1 = a_{n+2} + m$. We may view the inductive structure of type II again as a sort of "tree" beginning at dimension $n = 1$ and branching up through all dimensions, each step adding one more dimension.

One thing that makes the inductive structure especially useful is that the invariants we are dealing with do not change if we permute the entries in the character vector $\mathbf{a} = [a_0, a_1, \ldots, a_{n+1}]$. This means that all that we prove about induced characters is also true for characters that are "induced up to permutation". Up to permutation, a character may be induced from characters of lower dimension in many different ways, of course, and in particular may be *both* of type I *and* of type II.

Let $\mathcal{V} = \mathcal{V}_n^m(\mathbf{c})$ be a diagonal hypersurface of dimension n and degree m with twist $\mathbf{c}$ defined over $k = \mathbb{F}_q$ and let $\mathcal{X} = \mathcal{V}_n^m(\mathbf{1})$ be the corresponding Fermat variety. We fix m once and for all and vary n. In this section, we will be concerned with how the inductive structure is reflected in properties of the Hodge and Newton polygons of $\mathcal{V}$. Such properties are, as remarked above, independent of the twist $\mathbf{c}$. In other words, in this section we are essentially dealing with Fermat hypersurfaces $\mathcal{X}$.

THEOREM 4.1

Let $m \geq 3$ and $n \geq 1$ be as above. Then the following assertions hold:

(Type I) *Let $\mathbf{a} \in \mathfrak{A}_n^m$ and let $\mathcal{V}_A$ be the corresponding twisted Fermat motive. Then a twisted Fermat motive of dimension $n + 2$ of type I induced from $\mathbf{a}$ inherits the same structure as that of $\mathcal{V}_A$. In other words, if $\mathcal{V}_A$ is ordinary (resp. of Hodge–Witt type, resp. supersingular), then any induced motive of type I is ordinary (resp. of Hodge–Witt type, resp. supersingular).*

All $m - 1$ twisted Fermat motives of dimension $n + 2$ of type I induced from $\mathbf{a}$ are "cohomologically isomorphic" in the sense that they have the same cohomological invariants for Betti cohomology, ℓ-adic cohomology, crystalline cohomology, Hodge–Witt cohomology, etc.

More generally, any twisted Fermat motive of type I of dimension $n + 2d$ with $d \geq 1$ induced from **a** inherits the structure of $\mathcal{V}_A$.

(Type II) Let $\mathbf{a} \in \mathfrak{A}_{n+1}^m$ and $\mathbf{b} \in \mathfrak{A}_1^m$, and let $\mathcal{V}_A$ and $\mathcal{V}_B$ be the corresponding twisted Fermat motive of dimension $n+1$ and 1, respectively, where B denotes the $(\mathbb{Z}/m\mathbb{Z})^\times$-orbit of **b**. If both $\mathcal{V}_A$ and $\mathcal{V}_B$ are ordinary (resp. supersingular), then so is their induced motive of type II of dimension $n+2$. If $\mathcal{V}_A$ is of Hodge–Witt type and $\mathcal{V}_B$ is ordinary, or the other way around, then the induced motive is of Hodge–Witt type.

However, not all twisted Fermat motives of dimension $n + 2$ of type II induced from **a** are "cohomologically isomorphic".

More generally, for any positive integers r, s, let $\mathbf{a} \in \mathfrak{A}_r^m$ and $\mathbf{b} \in \mathfrak{A}_s^m$. If $\mathcal{V}_A$ and $\mathcal{V}_B$ are both ordinary (resp. both supersingular), then so is the induced motive of dimension $r + s$. If $\mathcal{V}_A$ is of Hodge–Witt type and $\mathcal{V}_B$ is ordinary, or the other way around, then $\mathcal{V}_{\tilde{A}}$ is of Hodge–Witt type.

The proof of Theorem 4.1 will be given bit-by-bit below, by looking into the effects of the inductive structure on the Newton and Hodge polygons.

We first set up the necessary notation. If $\tilde{\mathbf{a}} \in \mathfrak{A}_{n+2}^m$ is a character of type I (resp., type II) induced from $\mathbf{a} \in \mathfrak{A}_n^m$ (resp., $\mathbf{a} \in \mathfrak{A}_{n+1}^m$), let $\tilde{A}$ denote the $(\mathbb{Z}/m\mathbb{Z})^\times$-orbit of $\tilde{\mathbf{a}}$, and $\mathcal{V}_{\tilde{A}}$ the corresponding twisted Fermat motive of dimension $n + 2$ and degree m.

PROPOSITION 4.2
Let $m > 3$ and $n \geq 1$. Then the following assertions hold:

(Type I) Let $\mathbf{a} \in \mathfrak{A}_n^m$ and let $\mathcal{V}_A$ be the corresponding twisted Fermat motive of degree m and dimension n. Suppose that $\tilde{\mathbf{a}} \in \mathfrak{A}_{n+2}^m$ is an induced character of type I. Then the slopes of the Hodge polygon of the corresponding twisted Fermat motive $\mathcal{V}_{\tilde{A}}$ increase by 1 from those of $\mathcal{V}_A$ while keeping the same multiplicities. In other words, if $\mathcal{V}_A$ has a Hodge slope j, $0 \leq j \leq n$, with multiplicity h^j, then $\mathcal{V}_{\tilde{A}}$ has a Hodge slope $j + 1$ with multiplicity h^j.

More generally, if $\tilde{\mathbf{a}} \in \mathfrak{A}_{n+2d}^m$ is an induced character of type I, then the slopes of the Hodge polygon of the corresponding twisted Fermat motive increase by d from those of $\mathcal{V}_A$ while keeping the same multiplicities.

(Type II) Let $\mathbf{a} = (a_0, a_1, \ldots, a_{n+2}) \in \mathfrak{A}_{n+1}^m$, and $\mathbf{b} = (b_0, b_1, b_2) \in \mathfrak{A}_1^m$ with $a_{n+2} + b_2 = m$ and let $\mathcal{V}_A$ and $\mathcal{V}_B$ denote the corresponding twisted Fermat motives. Suppose that $\tilde{\mathbf{a}} = \mathbf{a} \# \mathbf{b} \in \mathfrak{A}_{n+2}^m$ is the induced character of type II. Then the slopes of the Hodge polygon of the corresponding twisted Fermat motive $\mathcal{V}_{\tilde{A}}$ are $\{\|t\mathbf{a}\| + \|t\mathbf{b}\|, t \in (\mathbb{Z}/m\mathbb{Z})^\times\}$, where $\|t\mathbf{b}\| \in \{0, 1\}$.

More generally, let $\tilde{\mathbf{a}} \in \mathfrak{A}_{r+s}^m$ be an induced character of type II, say, $\tilde{\mathbf{a}} = \mathbf{a} \# \mathbf{b}$ with $\mathbf{a} \in \mathfrak{A}_r^m$ and $\mathbf{b} \in \mathfrak{A}_s^m$. Then the slopes of the Hodge polygon of $\mathcal{V}_{\tilde{A}}$ are $\{\|t\mathbf{a}\| + \|t\mathbf{b}\|, t \in (\mathbb{Z}/m\mathbb{Z})^\times\}$, where $\|t\mathbf{b}\| \in \{0, 1, \ldots, s\}$.

Proof: We consider the two types in turn.

(Type I) Let $\tilde{\mathbf{a}} = (a_0, a_1, \ldots, a_{n+1}, a, m - a) \in \mathfrak{A}_{n+2}^m$ be an induced character of type I. Then

$$\|\tilde{\mathbf{a}}\| = \|\mathbf{a}\| + \langle \frac{a}{m} \rangle + \langle \frac{m-a}{m} \rangle = \|\mathbf{a}\| + 1.$$

It is easy to see that

$$\#\{\mathbf{a} \mid \|\mathbf{a}\| = j\} = h^j = \#\{\tilde{\mathbf{a}} \mid \|\tilde{\mathbf{a}}\| = j + 1\}.$$

(Type II) Let $\tilde{\mathbf{a}} = (a_0, a_1, \ldots, a_{n+1}, b_0, b_1)$ be an induced character of type II. Then

$$\|\tilde{\mathbf{a}}\| = \sum_{i=0}^{n+1} \langle \frac{a_i}{m} \rangle + \sum_{i=0}^{1} \langle \frac{b_i}{m} \rangle - 1 = \|\mathbf{a}\| + \|\mathbf{b}\| - (\langle \frac{a_{n+2}}{m} \rangle + \langle \frac{b_2}{m} \rangle - 1) = \|\mathbf{a}\| + \|\mathbf{b}\|. \quad \square$$

EXAMPLE 4.3 We give some examples of each type.

(Type I) Let $(m, n) = (5, 2)$. The set $\mathfrak{A}_2^5$ consists of 52 non-trivial characters. Grouping those characters which belong to isomorphic motives (i.e., group characters up to permutation and scaling), we obtain three different isomorphism classes: 16 characters are isomorphic to $\mathbf{a} = (1, 1, 1, 2)$, 24 are isomorphic to $\mathbf{a} = (1, 2, 3, 4)$ and 12 are isomorphic to $\mathbf{a} = (1, 1, 4, 4)$.

The Hodge polygon of $\mathcal{V}_{[1,1,1,2]}$ has slopes $0, 1, 2$ with multiplicities $1, 2, 1$, respectively. Then the Hodge polygons of the induced twisted Fermat motives of type I of dimensions $2 + 2d$ for any $d \geq 1$ have slopes $1 + d, 2 + d, 3 + d$ with the same multiplicity $1, 2, 1$.

The Hodge polygons of $\mathcal{V}_{[1,2,3,4]}$ and $\mathcal{V}_{[1,1,4,4]}$ have the pure slope 1 with multiplicity 4. Then the Hodge polygons of the induced twisted Fermat motives of type I dimensions $2 + 2d$ for any $d \geq 0$ have pure slope d with multiplicities 4.

Notice that $[1, 2, 3, 4]$ and $[1, 1, 4, 4]$ are both induced from the dimension zero character $[1, 4]$, so that one could obtain the information from the dimension zero case.

(Type II) Let $(m, n) = (7, n)$ with $n \geq 1$. Let $\mathbf{a} = (1, 1, 1, 4) \in \mathfrak{A}_2^7$. The Hodge polygon of $\mathcal{V}_{[1,1,1,4]}$ has slopes $0, 1, 2$ with respective multiplicities $2, 2, 2$. Now let $\mathbf{b} = (1, 3, 3) \in \mathfrak{A}_1^7$, and let $\tilde{\mathbf{a}} = \mathbf{a} \# \mathbf{b} = (1, 1, 1, 1, 3) \in \mathfrak{A}_3^7$ be an induced character of type II. Then the Hodge polygon of $\mathcal{V}_{[1,1,1,1,3]}$ has slopes $0, 1, 2, 3$ with multiplicities $1, 2, 2, 1$, respectively.

If $\mathbf{a} = (1, 1, 2, 4, 6) \in \mathfrak{A}_3^7$, then the Hodge polygon has slopes $1, 2$ with multiplicities $3, 3$, respectively. Let $\mathbf{b} = (1, 5, 1) \in \mathfrak{A}_1^7$ and let $\tilde{\mathbf{a}} = \mathbf{a} \# \mathbf{b} = (1, 1, 2, 4, 1, 5) \in \mathfrak{A}_4^7$ be an induced character of type II. Then the induced motive has the Hodge polygon with slopes $1, 2, 3$ with multiplicities $2, 2, 2$, respectively.

Now we shall discuss the effect of the inductive structures on Newton polygons.

PROPOSITION 4.4

Let $m > 3$ and let $n \geq 1$. Let p be a prime not dividing m and let f be the order of p mod m.

(Type I) Let $\mathbf{a} \in \mathfrak{A}_n^m$ and let $\mathcal{V}_A$ be the corresponding twisted Fermat motive of degree m and dimension n. Suppose that $\tilde{\mathbf{a}} \in \mathfrak{A}_{n+2}^m$ is an induced character of type I. Then the slopes of the Newton polygon of $\mathcal{V}_{\tilde{A}}$ increase by 1 from those of $\mathcal{V}_A$ while keeping the same multiplicities. In other words, if $\mathcal{V}_A$ has a Newton slope α with multiplicity r, then $\mathcal{V}_{\tilde{A}}$ has a Newton slope $\alpha + 1$ with multiplicity r.

More generally, if $\mathbf{a} \in \mathfrak{A}_{n+2d}^m$ is an induced character of type I from $\mathbf{a} \in \mathfrak{A}_n^m$, then the slopes of the Newton polygon of $\mathcal{V}_{\tilde{A}}$ increase by d from those of $\mathcal{V}_A$ while the multiplicities remain the same.

(Type II) If $\tilde{\mathbf{a}} \in \mathfrak{A}_{n+2}^m$ is of type II, say, $\tilde{\mathbf{a}} = \mathbf{a}\#\mathbf{b}$ where $\mathbf{a} \in \mathfrak{A}_{n+1}^m$ and $\mathbf{b} \in \mathfrak{A}_1^m$, then the slopes of $\mathcal{V}_{\tilde{A}}$ are $\{A_H(t\mathbf{a})/f + A_H(t\mathbf{b})/f, t \in (\mathbb{Z}/m\mathbb{Z})^\times\}$.

More generally, let $\tilde{\mathbf{a}} \in \mathfrak{A}_{r+s}^m$ be an induced character of type II, say, $\tilde{\mathbf{a}} = \mathbf{a}\#\mathbf{b}$ with $\mathbf{a} \in \mathfrak{A}_r^m$ and $\mathbf{b} \in \mathfrak{A}_s^m$. Then the slopes of the Newton polygon of $\mathcal{V}_{\tilde{A}}$ are $\{A_H(t\mathbf{a})/f + A_H(t\mathbf{b})/f, t \in (\mathbb{Z}/m\mathbb{Z})^\times\}$.

Proof: This follows from Lemma 1.3 and Lemma 2.3.

(Type I) Let $\mathbf{a} = (a_0, a_1, \ldots, a_{n+1}) \in \mathfrak{A}_n^m$ and let $\tilde{\mathbf{a}} = (a_0, a_1, \ldots, a_{n+1}, a, m - a) \in \mathfrak{A}_{n+2}^m$ be an induced character of type I. Then the slopes of the Newton polygon of $\mathcal{V}_{\tilde{A}}$ are given by $A_H(\tilde{\mathbf{a}})/f$ where

$$A_H(\tilde{\mathbf{a}}) = \sum_{t \in H} \|t\tilde{\mathbf{a}}\|.$$

But for each $t \in H$, we have

$$\|t\tilde{\mathbf{a}}\| = \|t\mathbf{a}\| + \langle \frac{a}{m} \rangle + \langle \frac{m-a}{m} \rangle = \|t\mathbf{a}\| + 1.$$

Hence the slopes of the Newton polygon of $\mathcal{V}_{\tilde{A}}$ are the slopes of $\mathcal{V}_A$ plus 1. The assertion on the multiplicity is obvious.

(Type II) Let $\tilde{\mathbf{a}} \in \mathfrak{A}_{n+2}^m$ be an induced character of type II, i.e., $\tilde{\mathbf{a}} = \mathbf{a}\#\mathbf{b}$ where $\mathbf{a} = (a_0, a_1, \ldots, a_{n+2})$ and $\mathbf{b} = (b_0, b_1, b_2)$ with $a_{n+2} + b_2 = m$. Then

$$\frac{A_H(\tilde{\mathbf{a}})}{f} = \frac{A_H(\mathbf{a}) + A_H(\mathbf{b})}{f} = \frac{A_H(\mathbf{a})}{f} + \frac{A_H(\mathbf{b})}{f}.$$

Hence the slopes of the Newton polygon of $\mathcal{V}_{\tilde{A}}$ are given by

$$\{A_H(t\mathbf{a})/f + A_H(t\mathbf{b})/f\}$$

as t ranges through $(\mathbb{Z}/m\mathbb{Z})^\times$.

EXAMPLE 4.5

(**Type I**) (1) Let $(m,n) = (7,d)$ with $d \geq 1$, and let p be a prime such that $p \equiv 2$ or $4 \pmod 7$. Then $f = 3$. The set $\mathfrak{A}_2^7$ consists of 186 characters. Up to permutation and scaling, they break up into isomorphism classes: 24 isomorphic to $\mathbf{a} = (1,1,1,4)$; 72 isomorphic to $\mathbf{a} = (1,1,2,3)$; 18 isomorphic to $\mathbf{a} = (1,1,6,6)$ and 12 isomorphic to $\mathbf{a} = (1,2,5,6)$.

The Newton polygon of $\mathcal{V}_{[1,1,1,4]}$ has slopes $1/3, 5/3$ with multiplicities $2,2$. Then the Newton polygon of the induced twisted Fermat motive of type I of dimension $2 + 2d$ for any $d \geq 0$ has slopes $1/3 + d, 5/3 + d$ with the same multiplicities $2,2$.

The Newton polygons of $\mathcal{V}_{[1,1,6,6]}$ and $\mathcal{V}_{[1,2,5,6]}$ have pure slope 1 with multiplicity 4. Then the Newton polygons of the induced twisted Fermat motives of type I dimension $2 + 2d$ for any $d \geq 0$ have pure slopes $1 + d$ with multiplicity 4.

(2) Let $(m,n) = (25,d)$ with $d \geq 1$. Let p be a prime such that $p \equiv 6$ or $21 \pmod{25}$. Then $f = 5$.

Let $\mathbf{a} = (1,1,5,18)$ or $\mathbf{a} = (1,3,5,16) \in \mathfrak{A}_{25}^2$. The Newton polygons $\mathcal{V}_A$ have slopes $3/5, 4/5, 6/5, 7/5$ with multiplicities $5,5,5,5$, respectively. Hence all the induced twisted Fermat motives of type I of dimension $2 + 2d$ have Newton polygons with slopes $3/5+d, 4/5+d, 6/5+d, 7/5+d$ with multiplicities $5,5,5,5$. Let $\mathbf{a} = (1,2,3,19) \in \mathfrak{A}_2^{25}$. The Newton polygon of $\mathcal{V}_A$ has the pure slope 1 with multiplicity 20. Hence all the induced twisted Fermat motives of type I of dimension $2 + 2d$ have Newton polygons with the pure slope $1 + d$ with multiplicity 20.

(**Type II**) Let $(m,n) = (7,n)$ with $n \geq 1$, and let p be a prime such that $p \equiv 2$ or $4 \pmod 7$, so that $f = 3$.

Let $\mathbf{a} = (1,1,1,2,2) \in \mathfrak{A}_3^7$. The Newton polygon of $\mathcal{V}_{[1,1,1,1,2]}$ has slopes $2/3, 7/3$ with multiplicities $3,3$. Let $\mathbf{b} = (1,1,5) \in \mathfrak{A}_1^7$ and let $\tilde{\mathbf{a}} = \mathbf{a}\#\mathbf{b} = (1,1,1,2,1,1) \in \mathfrak{A}_4^7$ be an induced character of type II. Then the Newton polygon of $\mathcal{V}_{\tilde{A}}$ has slopes $2/3 + 1/3, 7/3 + 2/3$ with multiplicities $3,3$.

Let $\mathbf{a} = (1,1,2,2,4,4) \in \mathfrak{A}_4^7$. The Newton polygon of $\mathcal{V}_{[1,1,2,2,4,4]}$ has slopes $1,3$ with multiplicities $3,3$. Let $\mathbf{b} = (b_0,b_1,3) \in \mathfrak{A}_1^7$ with $b_0+b_1 = 4$. Then there are two choices for $\mathbf{b}$, up to permutation, namely $\mathbf{b} = (1,3,3)$ and $(2,2,3)$.

Let $\tilde{\mathbf{a}} = \mathbf{a}\#(1,3,3) = (1,1,2,2,4,1,3) \in \mathfrak{A}_6^7$. Then the Newton polygon of the corresponding motive has slopes $5/3, 10/3$ with multiplicities $3,3$.

Let $\tilde{\mathbf{a}} = \mathbf{a}\#(2,2,3) \in \mathfrak{A}_6^7$. Then the Newton polygon of the corresponding motive has slopes $4/3, 11/3$ with multiplicities $3,3$.

PROPOSITION 4.6
Let $m > 3$ and $n \geq 1$. Then the following assertions hold:

(**Type I**) *Let $\mathbf{a} = (a_0,a_1,\ldots,a_{n+1}) \in \mathfrak{A}^m$ and let $\mathcal{V}_A$ be the corresponding twisted Fermat motive of degree m and dimension n. Suppose that $\tilde{\mathbf{a}} \in \mathfrak{A}_{n+2}^m$*

is an induced character of type I. Then $\mathcal{V}_{\tilde{A}}$ inherits the structure of $\mathcal{V}_A$, that is, if $\mathcal{V}_A$ is ordinary (resp. of Hodge–Witt type, resp. supersingular), then so is $\mathcal{V}_{\tilde{A}}$.

All $m-1$ twisted Fermat motives $\mathcal{V}_{\tilde{A}}$ of type I of dimension $n+2$ branching out from the same $\mathcal{V}_A$ of dimension n inherit the structure of $\mathcal{V}_A$.

(Type II) Let $\tilde{\mathbf{a}} = (a_0, a_1, \ldots, a_{n+1}, a_{n+2}) \in \mathfrak{A}_{n+1}^m$ and $\mathbf{b} = (b_0, b_1, b_2) \in \mathfrak{A}_1^m$ with $a_{n+2} + b_2 = m$. Suppose that $\tilde{\mathbf{a}} \in \mathfrak{A}_{n+2}^m$ is an induced character of type II, say, $\tilde{\mathbf{a}} = \mathbf{a}\#\mathbf{b}$. Let $\mathcal{V}_B$ denote the twisted Fermat motive corresponding to the $(\mathbb{Z}/m\mathbb{Z})^\times$-orbit B of $\mathbf{b}$. If both $\mathcal{V}_A$ and $\mathcal{V}_B$ are ordinary (resp. both are supersingular), then so is $\mathcal{V}_{\tilde{A}}$. If $\mathcal{V}_A$ is of Hodge–Witt type and $\mathcal{V}_B$ is ordinary, or the other way around, then $\mathcal{V}_{\tilde{A}}$ is of Hodge–Witt type.

However, not all twisted Fermat motives $\mathcal{V}_{\tilde{A}}$ of type II of dimension $n+2$ induced from the same $\mathcal{V}_A$ of dimension $n+1$ are "cohomologically isomorphic".

Proof: We use the combinatorial characterizations of ordinary motives, motives of Hodge–Witt type, and of supersingular motives given in Proposition 3.8.

(Type I) If $\mathcal{V}_A$ is ordinary (resp. of Hodge–Witt type), then for any $\mathbf{a} \in A$ and for any j, $0 \leq j < f$, we have

$$\|p^j\mathbf{a}\| - \|\mathbf{a}\| = 0 \quad (\text{resp.} \quad 0, \pm 1).$$

Then $\|p^j\tilde{\mathbf{a}}\| = \|p^j\mathbf{a}\| + 1$ for any j, $0 \leq j < f$, and moreover,

$$\|p^j\tilde{\mathbf{a}}\| - \|\tilde{\mathbf{a}}\| = \|p^j\mathbf{a}\| - \|\mathbf{a}\| = 0 \quad (\text{resp.} \quad 0, \pm 1).$$

This implies that $\mathcal{V}_{\tilde{A}}$ is ordinary (resp. of Hodge–Witt type). If $\mathcal{V}_A$ is supersingular, then $A_H(\mathbf{a}) = nf/2$ for any $\mathbf{a} \in A$, and hence $A_H(\tilde{\mathbf{a}})/f = A_H(\mathbf{a})/f+1 = (n+2)/2$. So $\mathcal{V}_{\tilde{A}}$ is also supersingular.

(Type II) Recall that
$$\|p^j\tilde{\mathbf{a}}\| = \|p^j\mathbf{a}\| + \|p^j\mathbf{b}\|$$

for any j, $0 \leq j \leq f$. Thus, if both $\mathcal{V}_A$ and $\mathcal{V}_B$ are ordinary, or if $\mathcal{V}_A$ is of Hodge–Witt type and $\mathcal{V}_B$ is ordinary, then $\|p^j\mathbf{b}\| = 0$ for any j, $0 \leq j \leq f$, so that

$$\|p^j\tilde{\mathbf{a}}\| - \|\tilde{\mathbf{a}}\| = 0 \quad \text{or} \quad 0, \pm 1 \quad \text{for any } j,\ 0 \leq j \leq f.$$

This implies that $\mathcal{V}_{\tilde{A}}$ is also ordinary, or of Hodge–Witt type. If both $\mathcal{V}_A$ and $\mathcal{V}_B$ are supersingular, then

$$A_H(\tilde{\mathbf{a}}) = A_H(\mathbf{a}) + A_H(\mathbf{b}) = fn/2 + f/2 = f(n+1)/2.$$

Therefore, $\mathcal{V}_{\tilde{A}}$ is supersingular. $\quad\square$

REMARK 4.7 For type II induced motives, the property "of Hodge–Witt type" will in general not be hereditary. Let $(m, n) = (7, 2)$ and let p be a prime such that $p \equiv 2$ or $4 \pmod 7$. Let $\mathbf{a} = (1, 1, 1, 4) \in \mathfrak{A}_2^7$ and $\mathbf{b} = (1, 3, 3) \in \mathfrak{A}_1^7$, and let $\tilde{\mathbf{a}} = \mathbf{a} \# \mathbf{b} = (1, 1, 1, 1, 3) \in \mathfrak{A}_3^7$. Then it is easy to see that $\mathcal{V}_A$ is of Hodge–Witt type and so is $\mathcal{V}_B$. However, for $\tilde{\mathbf{a}}$,

$$\|\mathbf{a}\| = 0, \quad \|2\mathbf{a}\| = 1, \quad \text{but} \quad \|4\mathbf{a}\| = 2.$$

This implies that $\mathcal{V}_{\tilde{A}}$ is not of Hodge–Witt type.

REMARK 4.8 For type II induction, notice that the requirement that a character of dimension 1 be supersingular forces f to be even, so that our result is non-empty only under that assumption. For a more powerful way (due to Shioda) of using the inductive structure to understand supersingular characters, see Appendix B and the references therein.

EXAMPLE 4.9

(Type I) (1) Let $(m, n) = (7, n)$ with $n \geq 1$. Let p be a prime such that $p \equiv 2$ or $4 \pmod 7$. Take $n = 2$. Then $\mathcal{V}_{[1,1,1,4]}$ is of Hodge–Witt type. Consequently, all the induced twisted Fermat motives $\mathcal{V}_{\tilde{A}}$ of type I of dimension $2 + 2d$ are of Hodge–Witt type, while the motive $\mathcal{V}_{[1,1,6,6]}$ is ordinary and supersingular. Therefore, all the twisted Fermat motives of type I stemming from this motive are also ordinary and supersingular.

(2) Let $(m, n) = (19, n)$ with $n \geq 1$. Let p be a prime such that $p \equiv 4$ or $5 \pmod{19}$. Take $n = 2$. Then $\mathcal{V}_{[1,4,5,9]}$ is of Hodge–Witt type. Therefore, all the induced twisted Fermat motives of dimension $2 + 2d$ are of Hodge–Witt type.

(Type II) Let $(m, n) = (7, n)$ with $n \geq 1$.

(1) Let p be a prime such that $p \equiv 2$ or $4 \pmod 7$. So $f = 3$. Let $\mathbf{a} = (1, 1, 2, 4, 6) \in \mathfrak{A}_3^7$. Then $\mathcal{V}_A$ is ordinary. Let $\mathbf{b} = (b_0, b_1, 1) \in \mathfrak{A}_1^7$ with $b_0 + b_1 = 6$. There are, up to permutation, three possible choices for $\mathbf{b}$, namely, $\mathbf{b} = (1, 5, 1)$, or $(2, 4, 1)$, or $(3, 3, 1)$. Let $\tilde{\mathbf{a}} = \mathbf{a} \# \mathbf{b} \in \mathfrak{A}_4^7$. If $\mathbf{b} = (2, 4, 1)$, $\mathcal{V}_B$ is ordinary and hence $\mathcal{V}_{\tilde{A}}$ is also ordinary. If $\mathbf{b} = (1, 5, 1)$ or $(3, 3, 1)$, $\mathcal{V}_B$ has the Hodge polygon with slopes $0, 1$ while the Newton polygon has slopes $1/3, 2/3$, so $\mathcal{V}_B$ is not ordinary. Consequently, $\mathcal{V}_{\tilde{A}}$ is not ordinary either.

(2) Let p be a prime such that $p \equiv 2$ or $4 \pmod 7$. So $f = 3$. Let $\mathbf{a} = (1, 1, 1, 5, 6) \in \mathfrak{A}_3^7$. Then $\mathcal{V}_A$ is of Hodge–Witt type but not ordinary. Let $\mathbf{b} = (b_0, b_1, 1) \in \mathfrak{A}_1^7$ with $b_0 + b_1 = 6$. Again, as in (1), there are three possibilities for $\mathbf{b}$. If $\mathbf{b} = (2, 4, 1)$, then $\mathcal{V}_B$ is ordinary. Now $\tilde{\mathbf{a}} = \mathbf{a} \# \mathbf{b} = (1, 1, 1, 5, b_0, b_1) \in \mathfrak{A}_4^7$ satisfies

$$\|p^j \tilde{\mathbf{a}}\| - \|\tilde{\mathbf{a}}\| = 0 \text{ or } \pm 1 \quad \text{for } j = 0, 1, 2.$$

Hence $\mathcal{V}_{\tilde{A}}$ is of Hodge–Witt type. If $\mathbf{b} = (1, 5, 1)$ or $(3, 3, 1)$, then $\mathcal{V}_B$ is of Hodge–Witt type. However, $\mathcal{V}_{\tilde{A}}$ is not of Hodge–Witt type as it violates the characterization of Proposition 3.8 for motives of Hodge–Witt type.

REMARK 4.10 Since characters in odd dimension are only supersingular when f is even, it is easy to find characters $\mathbf{a} \in \mathfrak{A}_{n+1}^m$ and $\mathbf{b} \in \mathfrak{A}_1^m$ such that neither $\mathcal{V}_A$ nor $\mathcal{V}_B$ are supersingular, but the induced character $\tilde{\mathbf{a}} = \mathbf{a} \# \mathbf{b}$ yields a supersingular motive $\mathcal{V}_{\tilde{A}}$. For example, let $m = 5$, $n = 2$, and choose any $p \equiv 1 \mod m$. Then neither $\mathbf{a} = [1, 1, 2, 3, 3] \in \mathfrak{A}_3^5$ nor $\mathbf{b} = [4, 4, 2] \in \mathfrak{A}_1^5$ are supersingular, but $\mathbf{a} \# \mathbf{b} = [1, 1, 2, 3, 4, 4] \in \mathfrak{A}_4^5$ certainly is.

To avoid the odd dimension problem, we can try to decompose into two characters of dimension two:

$$\mathbf{a} = [1, 1, 2, 3, 4, 4] = [1, 1, 2, 1] \# [3, 4, 4, 4],$$

and once again, neither of these characters in $\mathfrak{A}_2^5$ is supersingular. This gives a more convincing example that type II induction can produce supersingular characters from non-supersingular characters of smaller dimension.

For our specific example, things get better if we consider our character up to permutation. After a permutation, we get

$$\mathbf{a} = [1, 1, 2, 3, 4, 4] \sim [1, 2, 3, 4, 1, 4] = [1, 2, 3, 4] * [1, 4],$$

and both of these are supersingular.

The inductive structure can be realized also for formal groups arising from a twisted Fermat motive $\mathcal{V}_A$ and those arising from the induced motive $\mathcal{V}_{\tilde{A}}$. We denote by $\Phi_{\mathcal{V}_A}^{\bullet}$ the formal group arising from $\mathcal{V}_A$ and let $\mathcal{D}_{\mathcal{V}_A}^{\bullet}$ be its p-divisible part. For $\mathcal{V}_{\tilde{A}}$, $\Phi_{\mathcal{V}_{\tilde{A}}}^{\bullet}$ and $\mathcal{D}_{\mathcal{V}_{\tilde{A}}}^{\bullet}$ are defined similarly.

PROPOSITION 4.11
The hypotheses and the notation of Proposition 4.6 remain in force. Then the following assertions hold:

(Type I) *Let $\mathcal{V}_A$ be a twisted Fermat motive of dimension n and degree m. Let $\mathcal{V}_{\tilde{A}}$ be a type I twisted Fermat motive of dimension $n + 2$ induced from $\mathcal{V}_A$. Then for each i, $0 \leq i \leq n$, $\mathcal{D}_{\mathcal{V}_{\tilde{A}}}^{i+1, n+2-(i+1)}$ is isomorphic over $\bar{k}$ to $\mathcal{D}_{\mathcal{V}_A}^{i, n-i}$.*

(Type II) *Let $\mathcal{V}_A$ be a twisted Fermat motive of dimension $n + 1$ and degree m. Let $\mathcal{V}_{\tilde{A}}$ be a type II twisted Fermat motive of dimension $n + 2$ induced from $\mathcal{V}_A$ and $\mathcal{V}_B$. Then if $\mathcal{D}_{\mathcal{V}_A}^{i, n+1-i}$ is of multiplicative type and $\mathcal{V}_B$ is ordinary, then $\mathcal{D}^{i, n+2-i}$ is also of multiplicative type for each i, $0 \leq i \leq n + 1$.*

Proof:

(Type I) The slopes of $\mathcal{D}_{\mathcal{V}_A}^{i, n-i}$ coincide with those of $\mathcal{D}_{\mathcal{V}_{\tilde{A}}}^{i+1, n+2-(i+1)}$. In fact, the slopes of $\mathcal{D}_{\mathcal{V}_A}^{i,n-i}$ are given by

$$\{A_H(\mathbf{a})/f - i\}_{\mathbf{a} \in A} \quad \text{with} \quad i \le \frac{A_H(\mathbf{a})}{f} < i + 1;$$

while the slopes of $\mathcal{D}_{\mathcal{V}_{\tilde{A}}}^{i+1,n+2-(i+1)}$ are given by

$$\{A_H(\tilde{\mathbf{a}})/f - (i+1)\}_{\mathbf{a} \in \tilde{A}} \quad \text{with} \quad i+1 \le \frac{A_H(\tilde{\mathbf{a}})}{f} < i + 2.$$

But $A_H(\tilde{\mathbf{a}}) = A_H(\mathbf{a}) + 1$, so that slopes of these formal groups are equal. Therefore, the assertion follows, as over $\bar{k}$ slopes determine completely the structure of p-divisible groups.

(Type II) For each i, $0 \le i \le n+1$, slopes of $\mathcal{D}_{\tilde{A}}^{i,n+2-i}$ are given by

$$\left\{ \frac{A_H(\mathbf{a})}{f} - i + \frac{A_H(\mathbf{b})}{f} \right\},$$

where $\mathbf{a} \in A$ such that $i \le A_H(\mathbf{a})/f < i + 1$ by Lemma 3.5. If $\mathcal{V}_B$ is ordinary then $\Phi_{\mathcal{V}_B}^{i,1-i}$ is of multiplicative type for any I by Proposition 3.12. Hence the assertion follows from Proposition 4.6. $\square$

EXAMPLE 4.12 Let $(m,n) = (7,n)$ with $n \ge 1$. Let p be a prime such that $p \equiv 1 \pmod{7}$.

Let $\mathbf{a} = (1,1,1,4) \in \mathfrak{A}_2^7$. Then $\mathcal{D}_{\mathcal{V}_A}^{0,2}$ has slope 0 with multiplicity 2. Let $\tilde{\mathbf{a}} = (1,1,1,1,4,6) \in \mathfrak{A}_7^4$ be an induced character of type I. Then $\mathcal{D}_{\mathcal{V}_{\tilde{A}}}^{0,4}$ has slope 1 with multiplicity 2, while $\mathcal{D}_{\mathcal{V}_{\tilde{A}}}^{1,3}$ has slope 0 with multiplicity 2 and this is isomorphic to $\mathcal{D}_{\mathcal{V}_A}^{0,2}$ over $\bar{k}$.

Let $\tilde{\mathbf{a}} = \mathbf{a} \# \mathbf{b} = (1,1,1,3) \in \mathfrak{A}_3^7$ be an induced character of type II where $\mathbf{b} = (1,3,3) \in \mathfrak{A}_1^7$. Then $\mathcal{D}_{\mathcal{V}_{\tilde{A}}}^{0,3}$ has slope 0 with multiplicity 1.

Let $\mathbf{a} = (1,1,2,4,6) \in \mathfrak{A}_3^7$ and $\mathbf{b} = (2,4,1) \in \mathfrak{A}_1^7$. Then $\mathcal{V}_A$ and $\mathcal{V}_B$ are both ordinary. Let $\tilde{\mathbf{a}} = \mathbf{a} \# \mathbf{b} \in \mathfrak{A}_4^7$ be an induced character of type II. Then $\mathcal{D}_{\mathcal{V}_{\tilde{A}}}^{i,4-i}$ is isomorphic to a copy of $\hat{\mathbb{G}}_m$ over $\bar{k}$ for each $i, 0 \le i \le 4$.

REMARK 4.13 In the work of Ran [Ran81] and Shioda [Sh79a, Sh79b] on the Hodge conjecture for Fermat hypersurfaces, the inductive structure plays a crucial role. The central question is whether one can obtain any supersingular character of even dimension from *supersingular* characters of smaller even dimension. (We considered this question in a special case above, Remark 4.10.) When this is the case, one can use the inductive structure to construct cohomology classes that give a positive answer to the Hodge conjecture. Since one can describe supersingular characters in combinatorial terms, one can give a purely combinatorial characterization of the pairs (m,n) for which this process will work; Shioda calls this condition $P_m^n(p)$. One knows, however, that not every pair (m, n) satisfies Shioda's condition: in [Sh81], Shioda shows that for $m = 25$, $n = 4$, and $p \equiv 1 \pmod{m}$ there exist supersingular characters that do not come from supersingular characters of smaller dimension. See Appendix B for a few more details about this idea, and [Sh79a, Sh79b, Sh83b] for the full story.

5 Twisting and the Picard number

Let ℓ be a prime different from $p = \mathrm{char}(k)$.

For odd-dimensional diagonal hypersurfaces $\mathcal{V} = \mathcal{V}_n^m(\mathbf{c})$ with $n = 2d + 1$ over $k = \mathbb{F}_q$, the Tate conjecture is obviously true for any twist $\mathbf{c}$ as the ℓ-adic étale cohomology group $H^{2r}(\mathcal{V}_{\bar{k}}, \mathbb{Q}_\ell(r))$ has dimension 1 for any r, $0 \leq r \leq d$ (Milne [Mil86]).

Therefore, in this section, we confine ourselves to even-dimensional diagonal hypersurfaces $\mathcal{V} = \mathcal{V}_n^m(\mathbf{c})$ of dimension $n = 2d$ with twist $\mathbf{c}$ over $k = \mathbb{F}_q$. Let

$$Q^*(\mathcal{V}, T) := (1 - q^d T) \prod_{\mathbf{a} \in \mathfrak{A}_n^m} (1 - \mathcal{J}(\mathbf{c}, \mathbf{a}) T).$$

DEFINITION 5.1 *The d-th combinatorial Picard number $\rho_d(\mathcal{V}_k)$ is defined to be the multiplicity of q^d as a reciprocal root of the polynomial $Q^*(\mathcal{V}, T)$. That is,*

$$\rho_d(\mathcal{V}_k) = 1 + \#\mathfrak{B}_n^m$$

where

$$\mathfrak{B}_n^m = \{\mathbf{a} \in \mathfrak{A}_n^m \mid \mathcal{J}(\mathbf{c}, \mathbf{a}) = q^d\}.$$

We say that $\rho_d(\mathcal{V}_k)$ is stable if we have $\rho_d(\mathcal{V}_k) = \rho_d(\mathcal{V}_{k'})$ for any finite extension k' of k. There is always an extension k_1 such that $\rho_d(\mathcal{V}_{k_1})$ is stable. We write $\bar{\rho}_d(\mathcal{V})$ for the d-th stable combinatorial Picard number of $\mathcal{V}_k$.

One can think of the stable combinatorial Picard number of $\mathcal{V}$ as the combinatorial Picard number of the base change $\mathcal{V}_{\bar{k}}$ of $\mathcal{V}$ to the algebraic closure.

LEMMA 5.2
We have

$$\rho_d(\mathcal{V}_k) = 1 + \#\{\mathbf{a} \in \mathfrak{A}_n^m \mid j(\mathbf{a})/q^d = \chi(c_0^{a_0} c_1^{a_1} \dots c_{n+1}^{a_{n+1}})\}.$$

For the stable combinatorial Picard number, we have

$$\bar{\rho}_d(\mathcal{V}) = 1 + \#\overline{\mathfrak{B}}_n^m$$

where

$$\overline{\mathfrak{B}}_n^m = \{\mathbf{a} \in \mathfrak{A}_n^m \mid \mathcal{J}(\mathbf{c}, \mathbf{a})/q^d = \text{a root of unity in } L\}.$$

Proof: The first assertion is just the definition. The second follows at once from the Davenport–Hasse relations. □

Let $\mathcal{X} = V_n^m(1)$ denote the Fermat variety of dimension $n = 2d$ and degree m defined over k. We want to compare the numbers $\rho_d(\mathcal{X}_k)$, $\rho_d(\mathcal{V}_k)$, $\overline{\rho}_d(\mathcal{X})$ and $\overline{\rho}_d(\mathcal{V})$.

LEMMA 5.3
For the stable combinatorial Picard numbers, we have

$$\overline{\rho}_d(\mathcal{V}) = \overline{\rho}_d(\mathcal{X}).$$

Moreover, this quantity is equal to $1 + \sum B_n(V_{A_{\overline{k}}})$ where the sum is taken over all supersingular twisted Fermat motives $V_{A_{\overline{k}}}$.

Proof: The first assertion is clear, since over $\overline{k}$ we $\mathcal{V}_{\overline{k}} = \mathcal{X}_{\overline{k}}$. The second assertion follows immediately from Definition 5.1 and Lemma 5.2. □

Our computations are consistent with the following closed formulas for the stable d-th combinatorial Picard number $\overline{\rho}_d(\mathcal{V})$ of a diagonal hypersurface of dimension $2d$.

PROPOSITION 5.4
Assume that m is prime and that $p \equiv 1 \pmod{m}$, and let $\mathcal{V}$ be a diagonal hypersurface of dimension $n = 2d$ and of degree m. Then the following assertions hold:

1. *For $d = 1$, we have*

$$\overline{\rho}_1(\mathcal{V}) = 1 + (m-1)(3m - 6).$$

2. *For $d = 2$, we have*

$$\overline{\rho}_2(\mathcal{V}) = 1 + 5(m-1)(3m^2 - 15m + 20).$$

3. *For $d = 3$, we have*

$$\overline{\rho}_3(\mathcal{V}) = 1 + 5 \cdot 7(m-1)(3m^3 - 27m^2 + 86m - 95).$$

The statement for $d = 1$ is in fact a theorem proved by Shioda [Sh82a], who also indicates how such formulas can be obtained for higher dimensions, by a careful analysis of the inductive structure. For a discussion of the cases $d = 2$ and $d = 3$, see Appendix B. It is also worth pointing out that if we drop the condition that $p \equiv 1 \pmod{m}$ then these formulas still give lower bounds for the stable Picard number.

The following proposition gives a first result connecting combinatorial Picard numbers and stable combinatorial Picard numbers in the case when m is a prime number.

PROPOSITION 5.5
Assume that m is prime. Then the following assertions hold:

1. We have
$$\rho_d(\mathcal{X}_k) = \overline{\rho}_d(\mathcal{V}).$$
   That is, the actual d-th combinatorial Picard number of $\mathcal{X}_k$ is stable.

2. We have
$$\rho_d(\mathcal{V}_k) \leq \rho_d(\mathcal{X}_k).$$

3. *The following are equivalent:*

   (a) $\rho_d(\mathcal{V}_k) = \rho_d(\mathcal{X}_k)$,

   (b) $\mathbf{c}^{\mathbf{a}} = c_0^{a_0} c_1^{a_1} \ldots c_{n+1}^{a_{n+1}} \in (k^{\times})^m$ *for all supersingular* $\mathbf{a}$.

Proof: This is pretty much a direct consequence of Proposition 1.10. We know that $j(\mathbf{a}) = q^d$ for every supersingular character $\mathbf{a}$ (Lemma 3.7), which gives the first statement. The other two statements then follow immediately. $\square$

The condition that $\mathbf{c}^{\mathbf{a}}$ be an m-th power is closely connected, as in the proposition, with the variation of the combinatorial Picard number under twisting. We introduce some concepts intended to give a measure of this variation. For this discussion, we *assume that $m > 3$ is a prime throughout.* Recall that in this case we can only have $j(\mathbf{a}) = \xi q^d$ (ξ a root of unity) if $\xi = 1$, so that *for Fermat motives* "supersingular" and "strongly supersingular" are equivalent. The first important concern is to consider to what extent twisting preserves this property.

DEFINITION 5.6 *Suppose m is prime, $m > 3$. Let*
$$\mathcal{S} := \{\mathbf{a} \in \mathfrak{A}_n^m \mid j(\mathbf{a}) = q^d\} = \{\mathbf{a} \in \mathfrak{A}_n^m \mid \mathbf{a} \text{ is supersingular}\},$$
so that $\mathcal{S}$ is the set of supersingular $\mathbf{a}$'s.

Let $\mathbf{c} = (c_0, c_1, \cdots, c_{n+1})$ *be a twisting vector.*

1. *We say that $\mathbf{c}$ is* very mild *if $\mathbf{c}^{\mathbf{a}}$ is an m-th power for all $\mathbf{a} \in \mathcal{S}$.*

2. *We say that $\mathbf{c}$ is* extreme *if there is no $\mathbf{a} \in \mathcal{S}$ for which $\mathbf{c}^{\mathbf{a}}$ is an m-th power.*

The definitions are made so that the following assertions hold:

COROLLARY 5.7
If $m > 3$ is prime, we have

1. $\rho_d(\mathcal{V}_k) = \rho_d(\mathcal{X}_k)$ *whenever $\mathbf{c}$ is a very mild twist, and*

2. $\rho_d(\mathcal{V}_k) = 1$ *whenever $\mathbf{c}$ is an extreme twist.*

We would like to have some idea about how often these boundary cases occur. The first is in fact easy to decide:

PROPOSITION 5.8
The only very mild twist is the trivial twist.

Proof: As was pointed out above, the set $\mathcal{C}$ of all possible twisting vectors $\mathbf{c}$ (considered up to equivalence) is isomorphic to μ_m^{n+2}/Δ, where Δ is the diagonal inclusion of μ_m. Furthermore, we have a perfect pairing

$$\mathcal{C} \times \widehat{\mathfrak{G}} \longrightarrow \mu_m$$

where, as above,

$$\widehat{\mathfrak{G}} = \{(a_0, a_1, \ldots, a_{n+1}) \in (\mathbb{Z}/m\mathbb{Z})^{n+2} \mid \sum a_i = 0\},$$

mapping $(\mathbf{c}, \mathbf{a})$ to $\mathbf{c}^{\mathbf{a}}$. We can think of both $\mathcal{C}$ and $\widehat{\mathfrak{G}}$ as vector spaces over the field $\mathbb{F}$ with m elements. Recall, finally, that $\mathfrak{A} = \mathfrak{A}_n^m$ is the subset of $\widehat{\mathfrak{G}}$ given by the condition that $a_i \neq 0$ for all i.

A twist $\mathbf{c}$ is very mild if it annihilates every $\mathbf{a} \in \mathcal{S}$. If we denote by $\overline{\mathcal{S}}$ the vector subspace of $\widehat{\mathfrak{G}}$ generated by $\mathcal{S}$, it follows that $\mathbf{c}$ annihilates every vector in $\overline{\mathcal{S}}$. The proposition will follow then, from the following claim:

CLAIM: $\overline{\mathcal{S}} = \widehat{\mathfrak{G}}$.

To see this, note that the property of being supersingular is invariant under permutations of the entries a_i in the vector $\mathbf{a}$. From Lemma 5.9 below, this implies that $\overline{\mathcal{S}}$ (which is contained in $\widehat{\mathfrak{G}}$) is either trivial, one-dimensional, or equal to $\widehat{\mathfrak{G}}$. However, the estimate in Lemma 5.10 shows that the first two cases cannot occur, and we are done. $\square$

LEMMA 5.9
Let V be a finite dimensional vector space of dimension d over a field $\mathbb{F}$ of characteristic m. Let $W \subset V$ be a non-trivial subspace. Suppose that there exists a basis for V such that the action of the d-th symmetric group S_d given by permutation of the basis elements satisfies $\sigma(W) = W$ for all $\sigma \in S_d$. Then either $\operatorname{codim}(W) = 1$ or $\dim(W) = 1$.

Proof: Using the basis we have assumed exists, we may identify V with $\mathbb{F}^d$, and S_d acts by permuting the entries in a vector $(x_1, x_2, \ldots, x_d) \in V$. We may obviously assume $d \geq 4$, since the conclusion is trivially true otherwise.

The subspace
$$W_1 = \{(x, x, \ldots, x) \mid x \in \mathbb{F}\}$$
is then clearly the unique one-dimensional subspace which is fixed by all $\sigma \in S_d$. Dually, the subspace
$$W_0 = \{(x_1, x_2, \ldots, x_d) \mid \sum x_1 = 0\}$$

is also clearly the unique hyperplane in V which is invariant under all $\sigma \in S_d$. If m does not divide d, we clearly have $V = W_0 \oplus W_1$, and this direct sum decomposition is S_d-stable; on the other hand, if m does divide d, we have $W_1 \subset W_0$.

By hypothesis, W is a non-trivial subspace which is invariant under the action of S_d. We claim that we must have either $W = W_1$ or $W = W_0$. For this, assume that $W \neq W_1$, i.e., that there exists a vector $\mathbf{v} \in W$ whose entries are not all equal. We then proceed in several steps:

Step 1: there exists a vector $\mathbf{v}_1 \in W$ of the form $\mathbf{v}_1 = (0, x_2, x_3, \ldots, x_d)$.

We are assuming there is a vector $\mathbf{v} \in W$ not all of whose entries are equal. If any of those entries is equal to zero, we are done after a permutation. If they are all non-zero, let $\mathbf{v} = (y_1, y_2, y_3, \ldots, y_d) \in W$. Since W is closed under permutations, we have $(y_2, y_1, y_3, \ldots, y_d) \in W$, and hence

$$(y_1, y_2, y_3, \ldots, y_d) - \frac{y_1}{y_2}(y_2, y_1, y_3, \ldots, y_d) = (0, y_2 - \frac{y_1^2}{y_2}, (1 - \frac{y_1}{y_2})y_3, \ldots) \in W,$$

and this last vector is non-zero because $y_3 \neq 0$. This proves step 1.

Step 2: if W contains a vector of the form $(0, 0, \ldots, 0, 1, 1, \ldots, 1)$ consisting only of zeros and ones, then $W = W_0$ and m divides the number of ones in this vector.

Consider first the case when there is one zero and $d - 1$ ones. In this case, the space generated by W and the vector $(1, 1, \ldots, 1)$ must be all of V, since it contains

$$(1, 1, \ldots, 1) - (0, 1, \ldots, 1) = (1, 0, \ldots, 0)$$

and all its permutations. Since W is non-trivial, it must be of codimension 1 in V, and hence must be W_0 (which, as we pointed out above, is the unique S_d-invariant hyperplane).

Next, suppose

$$\underbrace{(0, 0, \ldots, 0, 1, 1, \ldots, 1)}_{i} \in W.$$

Let $\mathbf{v} \mapsto \tilde{\mathbf{v}}$ denote the projection on $\mathbb{F}^{d-i+1}$ given by the last $d - i + 1$ coordinates. The image of W under this projection is an S_{d-i+1}-invariant subspace of $\mathbb{F}^{d-i+1}$ which contains a vector consisting of one zero and $d - i$ ones. By the argument in the preceding paragraph, it must be the subspace defined by requiring that the sum of the entries be zero. Hence, there is a linear combination

$$\sum_j \lambda_j \sigma_j \tilde{\mathbf{v}} = (0, 0, \ldots, 0, 1, -1),$$

where $\lambda_j \in \mathbb{F}$ and $\sigma_j \in S_{d-i+1}$. Identifying S_{d-i+1} with a subgroup of S_d in the obvious way, it follows that "the same" linear combination works in W:

$$\sum_j \lambda_j \sigma_j \mathbf{v} = (0, 0, \ldots, 0, 1, -1).$$

Since $(0, 0, \ldots, 0, 1, -1)$ and its permutations clearly generate W_0, it follows that $W \supset W_0$, and hence $W = W_0$ because it is a non-trivial subspace.

The claim about the characteristic clearly follows from this conclusion. This proves step 2.

Step 3 (induction): repeat until done.

By step 1, we already know that W contains a non-zero vector of the form

$$(0, x_2, x_3, \ldots, x_d).$$

If all of the x_i are equal, we can divide by their common value and apply step 2 to conclude that $W = W_0$. If not, we can repeat step 1 as long as there are at least three non-zero entries.

Hence, we can conclude that either $W = W_0$ or W contains a non-zero vector with at most two non-zero entries. If there is only one non-zero entry, then clearly $W = V$, contrary to the hypothesis; hence there must be two. In addition, applying step 1 must yield the zero vector (because it cannot give a vector with only one non-zero entry). Hence, we must have either $(0, 0, \ldots, 0, 1, 1) \in W$, in which case step 2 applies, or $(0, 0, \ldots, 0, 1, -1) \in W$, in which case clearly $W \supset W_0$. This proves the lemma. $\square$

LEMMA 5.10
Let m be a prime, $m > 3$, and let $n = 2d$ be even. There exist at least

$$3(m-1)^d(m-2) + 1$$

supersingular characters $\mathbf{a} \in \mathfrak{A}_n^m$.

Proof: For $d = 1$, and $p \equiv 1 \pmod{m}$, this is true (with equality) by Proposition 5.4. If $d = 1$ and $p \not\equiv 1 \pmod{m}$, consider another prime p' such that $p' \equiv 1 \pmod{m}$. Then Proposition 5.4 applies to show that there are $3(m-1)(m-2) + 1$ supersingular characters when we take the base fields to be of characteristic p'. But we know that a character which is supersingular for p' is supersingular for any prime p. The inequality for $d = 1$ and general p follows.

Finally, for larger d, the inequality follows from the inductive structure, since each supersingular character $\mathbf{a}$ in dimension n yields $m - 1$ induced supersingular characters of type I in dimension $n + 2$ (see Theorem 4.1). (Of course, for $d = 2$ and $d = 3$ Proposition 5.4 gives much better lower bounds for $\#\mathcal{S}$.) $\square$

This allows a strengthening of Proposition 5.5:

THEOREM 5.11
Assume that m is prime. Then the following assertions hold:

1. *We have*
$$\rho_d(\mathcal{X}_k) = \overline{\rho}_d(\mathcal{V}).$$
   That is, the actual d-th combinatorial Picard number of $\mathcal{X}_k$ is stable.

2. *We have*
$$\rho_d(\mathcal{V}_k) \le \rho_d(\mathcal{X}_k).$$

3. *The following are equivalent:*

   (a) *$\mathcal{V}_k$ and $\mathcal{X}_k$ are isomorphic,*

   (b) *$\rho_d(\mathcal{V}_k) = \rho_d(\mathcal{X}_k)$,*

   (c) *$\mathbf{c}^{\mathbf{a}} = c_0^{a_0} c_1^{a_1} \dots c_{n+1}^{a_{n+1}} \in (k^{\times})^m$ for all supersingular $\mathbf{a}$,*

   (d) *$\mathbf{c}$ is equivalent to the trivial twist.*

Proof: The first two assertions are from Proposition 5.5, so the issue is proving the third statement. For that, note first that it is clear that isomorphic varieties will have the same combinatorial Picard number, so (3a) implies (3b). The equivalence between (3b), (3c), and (3d) follows at once from Proposition 5.5 and Proposition 5.8, and (3d) clearly implies (3a). □

Understanding extreme twists is much harder. Note, first, that they do exist: if $\mathbf{c} = (c_0, 1, 1, \dots, 1)$ with $c_0 \ne 1$, then $\mathbf{c}^{\mathbf{a}} = c_0^{a_0}$ cannot be an m-th power unless c_0 is already an m-th power (since m is prime and $a \not\equiv 0$ (mod m)). Hence, if c_0 is not an m-th power in k the twist $\mathbf{c} = (c_0, 1, 1, \dots, 1)$ is extreme. In our computations, *all* extreme twists turn out to be equivalent to twists of this form. Hence, it is natural to ask:

Question: How many extreme twists are there for a given choice of m, n, and k? Are all of them, up to equivalence, of the form $\mathbf{c} = (c_0, 1, 1, \dots, 1)$ with $c_0 \ne 1$?

We now proceed to relate this combinatorial game with matters of more serious import:

DEFINITION 5.12 *Let $\mathcal{V} = \mathcal{V}_n^m(\mathbf{c})$ be a diagonal hypersurface of degree m, dimension $n = 2d$ with twist $\mathbf{c}$ over $k = \mathbb{F}_q$. The d-th ℓ-adic Picard number, $\rho_{d,\ell}(\mathcal{V}_k)$, of $\mathcal{V}_k$ is defined to be the dimension of the ℓ-adic étale cohomology group $H^{2d}(\mathcal{V}_{\overline{k}}, \mathbb{Q}_\ell(d))$, generated by algebraic cycles of codimension d on $\mathcal{V}$ over k.*

CONJECTURE 5.13 (THE TATE CONJECTURE) *With notation as above, we define the d-th Picard number of* $\mathcal{V}_k$ *by*

$$\rho_d'(\mathcal{V}_k) := \max_{\ell \neq p} \rho_{d,\ell}(\mathcal{V}_k),$$

where ℓ *runs over all primes not equal to p. Then* $\rho_d'(\mathcal{V}_k)$ *is a well-defined quantity, which is equal to the rank of the Chow group of algebraic cycles of codimension d on* $\mathcal{V}_k$ *modulo rational equivalence, and*

$$\rho_d'(\mathcal{V}_k) = \rho_d(\mathcal{V}_k).$$

(As we pointed out before, it is known that $\rho_d'(\mathcal{V}_k) \leq \rho_d(\mathcal{V}_k)$. So the Tate conjecture claims the validity of the reverse inequality.)

DEFINITION 5.14 *We say a diagonal hypersurface* $\mathcal{V} = \mathcal{V}_n^m(\mathbf{c})$ *over k is extreme if the d-th Picard number* $\rho_d'(\mathcal{V}_k) = 1$.

(Observe that if $\mathcal{V}_k$ is strongly supersingular over k, it can never be extreme.)

THEOREM 5.15
Let $\mathcal{V} = \mathcal{V}_n^m(\mathbf{c})$ *be a diagonal hypersurface of prime degree* $m > 3$ *and dimension* $n = 2d$ *with twist* $\mathbf{c}$ *over* $k = \mathbb{F}_q$. *If* $\mathbf{c}$ *is extreme, then* $\mathcal{V}$ *is extreme, and in this case the Tate conjecture holds for* $\mathcal{V}_k$ *and we have*

$$\rho_d'(\mathcal{V}_k) = \rho_d(\mathcal{V}_k) = 1.$$

Proof: If $\mathbf{c}$ is extreme, then $\rho_d(\mathcal{V}_k) = 1$ by Corollary 5.7. Since there is always an obvious algebraic cycle of codimension d on $\mathcal{V}$ defined over k, namely, the algebraic cycles of hyperplane sections of codimension d on $\mathcal{V}_k$, the assertion follows from the inequality $\rho_d'(\mathcal{V}_k) \leq \rho_d(\mathcal{V}_k)$. □

REMARK 5.16 Shioda [Sh83a] constructed an example of hypersurfaces Y, of degree m and dimension n in $\mathbb{P}_{\mathbb{F}_p}^{n+1}$, having Picard number $\rho_{n/2}(Y) = 1$. Our examples are different from those of Shioda.

For diagonal hypersurfaces $\mathcal{V} = \mathcal{V}_n^m(\mathbf{c})$ over $k = \mathbb{F}_q$ with non-extreme twists $\mathbf{c}$, the following results on the Tate conjecture follow from results of Tate [Ta65], Shioda [Sh79a, Sh79b], and Ran [Ran81].

PROPOSITION 5.17
Let $\mathcal{V} = \mathcal{V}_n^m(\mathbf{c})$ *be a diagonal hypersurface of dimension* $n = 2d$ *and degree* m *with twist* $\mathbf{c}$ *over* $k = \mathbb{F}_q$. *Then the Tate conjecture holds for* $\mathcal{V}_k$ *in the following cases:*

1. $n = 2$ and m, $\mathbf{c}$ and p arbitrary;

2. $m > 3$ is prime, n is arbitrary, and $p \equiv 1 \pmod{m}$;

3. $m \leq 20$, n is arbitrary, and $p \equiv 1 \pmod{m}$;

4. $m = 21$, $n \leq 10$ and $p \equiv 1 \pmod{m}$;

5. $\mathcal{V}_k$ is supersingular.

Proof: First of all, notice that if the Tate conjecture holds for $\mathcal{X}_{\bar{k}}$ then it holds for any diagonal hypersurface $\mathcal{V}$ such that $\mathcal{V}_{\bar{k}} \cong \mathcal{X}_{\bar{k}}$. To see this, note that if the Tate conjecture holds over $\bar{k}$, then $H^n(\mathcal{V}_{\bar{k}}, \mathbb{Q}_\ell(d)) = H^n(\mathcal{X}_{\bar{k}}, \mathbb{Q}_\ell(d))$ is spanned by algebraic cycles (of codimension d) defined over $\bar{k}$. Taking the $\mathrm{Gal}(\bar{k}/k)$-invariant subspace of this $\mathbb{Q}_\ell$ vector space, we see that $H^n(\mathcal{V}_{\bar{k}}, \mathbb{Q}_\ell(d))^{\mathrm{Gal}(\bar{k}/k)}$ is also spanned by algebraic cycles (of codimension d) defined over k, and hence the Tate conjecture holds for $\mathcal{V}$.

Hence, we can work over the algebraic closure, in which case it suffices to prove the conjecture for Fermat hypersurfaces. Shioda and Katsura [SK79] show that the Tate conjecture holds for Fermat hypersurfaces of dimension $n = 2$ and arbitrary m and p, and Shioda [Sh79a, Sh79b] shows that it holds in all but the last case we enumerate above. Finally, by Tate [Ta65], the Tate conjecture holds for a supersingular Fermat variety $\mathcal{X}_{\bar{k}}$, which concludes the proof. $\square$

Shioda [Sh79a, Sh79b] shows that the Tate conjecture holds whenever certain combinatorial conditions $P^n_m(p)$ hold. The list of cases above corresponds to situations where one has been able to verify these conditions. The conditions are quite explicit, and can easily be checked computationally for any given m, n, and p. It is known that they do not always hold; in [Sh81], a counterexample is given for $m = 25$, $n = 4$, and $p \equiv 1 \pmod{m}$. For more on the case when m is prime, see Appendix B.

Picard numbers for when m is composite. All our results in this chapter have assumed that m is prime; it is natural to ask what happens when this assumption is relaxed. Unfortunately, there is very little we can say.

Assume that $\mathcal{V}_k$ has composite degree. In this case, we observe that the stable and the actual combinatorial Picard numbers are considerably different from the prime degree case. For instance, Proposition 5.5 (2) no longer holds: in some cases one can find twists **c** satisfying

$$\rho_d(\mathcal{V}_k) > \rho_d(\mathcal{X}_k),$$

where

$$\rho_d(\mathcal{V}_k) = \#\{\mathbf{a} \in \mathfrak{A}^m_n \mid \chi(\mathbf{c}^{\mathbf{a}}) = j(\mathbf{a})/q^d\}.$$

One can also find non-trivial twists for which $\rho_d(\mathcal{V}_k) = \rho_d(\mathcal{X}_k)$, in contrast to what holds for prime degree (Theorem 5.11).

The situation is less confusing when m is odd. See Section 9.1 for some speculations about this case, and the tables in Section A.3 for some data.

Tables of Picard numbers In the appendix, we list examples of the combinatorial Picard numbers $\rho_d(\mathcal{X}_k)$, $\rho_d(\mathcal{V}_k)$ and of the stable combinatorial Picard numbers $\overline{\rho}_d(\mathcal{V})$ of $\mathcal{V} = \mathcal{V}_n^m$ of dimension $n = 2d$ with twists $\mathbf{c}$ defined over the prime field $k = \mathbb{F}_p$. (In many of the cases covered by the tables the Tate conjecture is known to hold by Proposition 5.17, so that we are obtaining the geometric Picard number.) The results are presented in Section A.3.

6 "Brauer numbers" of twisted Fermat motives

Let $\mathcal{V} = \mathcal{V}_n^m(\mathbf{c})$ be a diagonal hypersurface of dimension n and degree m with twist $\mathbf{c}$ defined over $k = \mathbb{F}_q$. The evaluation of the polynomials $Q^*(\mathcal{V}, T)$ (for $n = 2d$) and $Q(\mathcal{V}, T)$ (for $n = 2d+1$) at $T = q^{-r}$ for each integer r, $0 \le r \le d$, can be reduced to the evaluation of the polynomials $Q(\mathcal{V}_A, T)$ and hence to the computation of the norms of the form

$$\mathrm{Norm}_{L/\mathbb{Q}}(1 - \frac{\mathfrak{J}(\mathbf{c}, \mathbf{a})}{q^r}) = Q(\mathcal{V}_A, q^{-r})$$

for each integer r, $0 \le r \le d$.

We first recall some relevant notation. If ℓ is any prime, let $| \ |_\ell^{-1}$ denote the ℓ-adic valuation normalized by $|\ell|_\ell^{-1} = \ell$. For the prime $p = \mathrm{char}(k)$ and $k = \mathbb{F}_q$, let ν denote a p-adic valuation normalized by $\nu(q) = 1$.

LEMMA 6.1
Let $\mathcal{V}_A$ be a twisted Fermat motive of dimension n and of degree $m = m_0^t$, where m_0 is an odd prime and $m > 3$, with twist $\mathbf{c}$ over $k = \mathbb{F}_q$. Suppose that $\mathcal{V}_A$ is supersingular. Then for any r, $0 \le r \le n$,

$$\mathrm{Norm}_{L/\mathbb{Q}}(1 - \frac{\mathfrak{J}(\mathbf{c}, \mathbf{a})}{q^r}) = \prod_{t \in (\mathbb{Z}/m\mathbb{Z})^\times} (1 - q^{\frac{n}{2} - r} \xi^t)$$

where ξ is some m-th root of unity in L.

In particular, if $\mathcal{V}_A$ is strongly supersingular, then for any r, $0 \le r \le n$,

$$\mathrm{Norm}_{L/\mathbb{Q}}(1 - \frac{\mathfrak{J}(\mathbf{c}, \mathbf{a})}{q^r}) = (1 - q^{\frac{n}{2} - r})^{B_n(\mathcal{V}_A)}.$$

If $2r = n$ and $\mathcal{V}_A$ is not strongly supersingular, the norm is divisible by m_0. In particular, if m is prime and $\mathbf{c}$ has the property that $\mathbf{c^a} \notin (k^\times)^m$ for $\mathbf{a} \in A$, then the norm is equal to m.

Proof: If $\mathcal{V}_A$ is supersingular, then $\mathfrak{J}(\mathbf{c}, \mathbf{a}) = q^{\frac{n}{2}} \xi$ for some m-th root of unity $\xi \in L$. If $2r = n$, then the norm is equal to $\prod_{t \in (\mathbb{Z}/m\mathbb{Z})^\times} (1 - \xi^t)$, which is

61

obviously divisible by m_0. If m is prime, $\mathbf{c^a} \notin (k^\times)^m$ for some (and hence any) $\mathbf{a} \in A$, then the norm is exactly equal to m by Proposition 1.10. $\quad\square$

Note that if $\mathcal{V}_A$ is strongly supersingular and $2r = n$ the norm is simply equal to zero.

THEOREM 6.2

Let $\mathcal{V}_A$ be a twisted Fermat motive of dimension n and prime degree $m > 3$ with twist $\mathbf{c}$ over $k = \mathbb{F}_q$. Suppose that $\mathcal{V}_A$ is of Hodge–Witt type and is not supersingular. Then the following assertions hold.

1. If $n = 2d$, then for any r, $0 \le r \le n$,

$$\text{Norm}_{L/\mathbb{Q}}(1 - \frac{\mathfrak{J}(\mathbf{c}, \mathbf{a})}{q^r}) = \frac{B^r(\mathcal{V}_A) \cdot m}{q^{w(r)}}$$

where

$$w(r) := w_{\mathcal{V}_A}(r) = \sum_{i=0}^{r} (r - i) h^{i, n-i}(\mathcal{V}_A),$$

and $B^r(\mathcal{V}_A)$ is a positive integer (not necessarily prime to mp) satisfying

$$B^r(\mathcal{V}_A) = B^{n-r}(\mathcal{V}_A).$$

2. If $n = 2d + 1$, then for any r, $0 \le r \le n$,

$$\text{Norm}_{L/\mathbb{Q}}(1 - \frac{\mathfrak{J}(\mathbf{c}, \mathbf{a})}{q^r}) = \frac{D^r(\mathcal{V}_A)}{q^{w(r)}}$$

where $D^r(\mathcal{V}_A)$ is a positive integer (not necessarily prime to mp) satisfying

$$D^r(\mathcal{V}_A) = D^{n-r}(\mathcal{V}_A),$$

and $w(r)$ is the integer defined in (1).

Proof: Our proof will be divided into several parts. (Cf. Shioda [Sh87], Suwa and Yui [SY88], Suwa [Su91a, Su91b], and also Yui [Yu91].)

(1) The p-part: By Lemma 3.5 (3), we know that the set of slopes of $\mathcal{V}_A$ is $\{A_H(\mathbf{a})/f \mid \mathbf{a} \in A\}$. Thus, the exponent of q in the norm is given by

$$- \sum_{\mathbf{a} \in A} \max(0, r - A_H(\mathbf{a})/f) = - \sum_{i=0}^{r} (r - i) \#\{\mathbf{a} \in A \mid A_H(\mathbf{a}) = if\}.$$

If $\mathcal{V}_A$ is ordinary, then by Proposition 3.8 (1), we have

$$\#\{\mathbf{a} \in A \mid A_H(\mathbf{a}) = if\} = \#\{\mathbf{a} \in A \mid \|\mathbf{a}\| = i\} = h^{i, n-i}(\mathcal{V}_A).$$

If $\mathcal{V}_A$ is of Hodge–Witt type, then again by Proposition 3.8 (3), $\|p^j\mathbf{a}\| - \|\mathbf{a}\|$ takes values 0 or ± 1. If $\|p^j\mathbf{a}\| = \|\mathbf{a}\|$ for every j, $0 \le j < n$, then $\mathcal{V}_A$ is ordinary. So we assume that $\|p^j\mathbf{a}\| = \|\mathbf{a}\| + 1$ for some j, and for each i, $0 \le i \le n$, define the following quantities:

$$s_{-1,i} = \#\{j, 0 \le j < f \mid \|\mathbf{a}\| = i \quad \text{and} \quad \|p^j\mathbf{a}\| = i - 1\},$$
$$s_{0,i} = \#\{j, 0 \le j < f \mid \|\mathbf{a}\| = i \quad \text{and} \quad \|p^j\mathbf{a}\| = i\},$$
$$s_{1,i} = \#\{j, 0 \le j < f \mid \|\mathbf{a}\| = i \quad \text{and} \quad \|p^j\mathbf{a}\| = i + 1\}.$$

Then for each i, $0 \le i \le n$, $s_{-1,i} + s_{0,i} + s_{1,i} = f$. Furthermore, for each $\mathbf{a} \in A$ with $\|\mathbf{a}\| = i$, $0 \le i \le n$, we have

$$r - \frac{A_H(\mathbf{a})}{f} = (r - i) - \frac{1}{f}(\|\mathbf{a}\| + \|p\mathbf{a}\| + \cdots + \|p^j\mathbf{a}\| + \cdots + \|p^{f-1}\mathbf{a}\|)$$

$$= r - \frac{1}{f}\{if + (s_{1,i} - s_{-1,i})\} = (r - i) - \frac{1}{f}(s_{1,i} - s_{-1,i}).$$

Then the exponent of q in the norm is given by

$$-\sum_{i=0}^{r}(r - i)h^{i,n-i}(\mathcal{V}_A) + \frac{1}{f}\sum_{i=0}^{r}(s_{1,i} - s_{-1,i})h^{i,n-i}(\mathcal{V}_A).$$

But by the duality, we have for each i,

$$h^{i,n-i}(\mathcal{V}_A) = h^{n-i,i}(\mathcal{V}_A) \quad \text{and} \quad (s_{1,i} - s_{-1,i}) + (s_{1,n-i} - s_{-1,n-i}) = 0.$$

This implies that the second sum in the above expression is 0, and therefore the exponent of q in the norm is

$$-\sum_{i=0}^{r}(r - i)h^{i,n-i}(\mathcal{V}_A) = -w(r).$$

(2) The m-part: This follows from the fact that $\mathfrak{J}(\mathbf{c}, \mathbf{a}) \equiv 1 \pmod{(1 - \zeta)}$ (Proposition 1.7).

The proofs of the remaining assertions, that, for each r, $0 \le r \le n$, we have $B^r(\mathcal{V}_A) = B^{n-r}(\mathcal{V}_A)$ (resp. $D^r(\mathcal{V}_A) = D^{n-r}(\mathcal{V}_A)$), are deferred to Proposition 6.4 and its corollary below. $\square$

REMARK 6.3

1. In part (1) of Theorem 6.2, if we take a twist $\mathbf{c}$ satisfying $\mathbf{c}^{\mathbf{a}} \notin (k^\times)^m$ for any $\mathbf{a} \in A$, then the exponent of m in the norm $\mathrm{Norm}_{L/\mathbb{Q}}(1 - \mathfrak{J}(\mathbf{c}, \mathbf{a})/q^d)$ is precisely equal to 1. In this case, therefore, $B^d(\mathcal{V}_A)$ is relatively prime to m. This follows from Proposition 1.7 and Proposition 1.10.

2. In part (2) of Theorem 6.2, if we ease the assumption that $\mathcal{V}_A$ is of Hodge–Witt type, the invariant $w(r)$ ought to be adjusted. That is, $w(r)$ should be replaced by

$$w(r) - T^{r-1,n-r+1}(\mathcal{V}_A)$$

where $T^{r-1,n-r+1}(\mathcal{V}_A)$ is defined to be the dimension of the unipotent formal group $\mathcal{U}^{n+1}(\mathcal{V}_A, \mathbb{Z}_p(r))$ (cf. Suwa and Yui [SY88], Remark (II.2.3)). Some results in this direction have just been announced by Suwa [Su].

PROPOSITION 6.4 (MILNE [MIL86], §10)
There is a duality between the norms. That is, for each r, $0 \le r \le [n/2]$,

$$\frac{\mathrm{Norm}_{L/\mathbb{Q}}(1 - \mathfrak{J}(\mathbf{c},\mathbf{a})/q^r)}{\mathrm{Norm}_{L/\mathbb{Q}}(1 - \mathfrak{J}(\mathbf{c},\mathbf{a})/q^{n-r})} = q^{w(n-r)-w(r)}.$$

Proof: This is essentially the result of Milne in §10 of [Mil86], projected down to motives. We sketch a proof here. We know (Weil [We49, We52]) that for any r, $0 \le r \le n$, V satisfies the functional equation

$$\lim_{s \to r} \frac{Z(V, q^{(n-s)})}{Z(V, q^{-s})} = q^{(\frac{n}{2}-r)\mathcal{E}}$$

where $\mathcal{E} = \sum_{i=0}^{2n}(-1)^i B_i(V)$ is the self-intersection number of the diagonal $V \times V$ (which is the Euler Poincaré characteristic of V). Now the functional equation commutes with the motivic decomposition, so we obtain, for each i, $0 \le i \le n$,

$$\frac{Q(\mathcal{V}_A, q^{-r})}{Q(\mathcal{V}_A, q^{n-r})} = q^{(r-\frac{n}{2})B_n(\mathcal{V}_A)}.$$

But $B_n(\mathcal{V}_A) = \sum_{i=0}^n h^{i,n-i}(\mathcal{V}_A)$ by Lemma 2.3. So by applying the same argument as in [Mil86], (10.1), we have

$$(r - \frac{n}{2})B_n(\mathcal{V}_A) = \sum_{i=0}^n (n-r-i)h^{i,n-i}(\mathcal{V}_A) - \sum_{i=0}^n (r-i)h^{i,n-i}(\mathcal{V}_A)$$

$$= w(n-r) - w(r). \quad \square$$

COROLLARY 6.5
For each r, $0 \le r \le n$,

$$B^r(\mathcal{V}_A) = B^{n-r}(\mathcal{V}_A) \qquad \text{and} \qquad D^r(\mathcal{V}_A) = D^{n-r}(\mathcal{V}_A).$$

Proof: Note that for each i, $0 \le i \le n$,

$$\frac{Q(\mathcal{V}_A, q^{-r}) \cdot q^{w(n-r)}}{Q(\mathcal{V}_A, q^{n-r}) \cdot q^{w(r)}} = 1. \qquad \square$$

EXAMPLE 6.6 Here are some examples. Let $(m,n) = (5,4)$ and let $\mathbf{c} = (1,1,1,1,1,3)$. Let $q = p \in \{11, 31, 41\}$. We compute the norms for $1 \leq r \leq 3$ and different vectors $\mathbf{a}$:

1. Let $\mathbf{a} = (1,1,1,1,2,4)$. Then $w(1) = 0, w(2) = 1$ and $w(3) = 4$.

$$\text{Norm}_{L/\mathbb{Q}}(1 - \mathcal{J}(\mathbf{c}, \mathbf{a})/11^r) = \begin{cases} 5 \cdot 41 \cdot 71 & \text{for} \quad r = 1 \\ 5/11 & \text{for} \quad r = 2 \\ 5 \cdot 41 \cdot 71/11^4 & \text{for} \quad r = 3. \end{cases}$$

$$\text{Norm}_{L/\mathbb{Q}}(1 - \mathcal{J}(\mathbf{c}, \mathbf{a})/31^r) = \begin{cases} 5 \cdot 271 \cdot 661 & \text{for} \quad r = 1 \\ 5/31 & \text{for} \quad r = 2 \\ 5 \cdot 271 \cdot 661/31^4 & \text{for} \quad r = 3. \end{cases}$$

$$\text{Norm}_{L/\mathbb{Q}}(1 - \mathcal{J}(\mathbf{c}, \mathbf{a})/41^r) = \begin{cases} 3^4 \cdot 5^3 \cdot 281 & \text{for} \quad r = 1 \\ 5^3/41 & \text{for} \quad r = 2 \\ 3^4 \cdot 5^3 \cdot 281/41^4 & \text{for} \quad r = 3. \end{cases}$$

2. Let $\mathbf{a} = (1,1,1,1,3,3)$. Then $w(1) = 0, w(2) = 2$, and $w(3) = 4$.

$$\text{Norm}_{L/\mathbb{Q}}(1 - \mathcal{J}(\mathbf{c}, \mathbf{a})/11^r) = \begin{cases} 5 \cdot 11 \cdot 271 & \text{for} \quad r = 1 \\ 3^4 \cdot 5/11^2 & \text{for} \quad r = 2 \\ 5 \cdot 11 \cdot 271/11^4 & \text{for} \quad r = 3. \end{cases}$$

$$\text{Norm}_{L/\mathbb{Q}}(1 - \mathcal{J}(\mathbf{c}, \mathbf{a})/31^r) = \begin{cases} 3^4 \cdot 5 \cdot 11 \cdot 191 & \text{for} \quad r = 1 \\ 3^4 \cdot 5/31^2 & \text{for} \quad r = 2 \\ 3^4 \cdot 5 \cdot 11 \cdot 191/31^4 & \text{for} \quad r = 3. \end{cases}$$

$$\text{Norm}_{L/\mathbb{Q}}(1 - \mathcal{J}(\mathbf{c}, \mathbf{a})/41^r) = \begin{cases} 5^3 \cdot 11 \cdot 2131 & \text{for} \quad r = 1 \\ 3^4 \cdot 5^3/41^2 & \text{for} \quad r = 2 \\ 5^3 \cdot 11 \cdot 2131/41^4 & \text{for} \quad r = 3. \end{cases}$$

Cohomological interpretations of the integers $B^r(\mathcal{V}_A)$ and $D^r(\mathcal{V}_A)$ for any r, $0 \leq r \leq n$ are given as follows. Let $\Gamma = \text{Gal}(\bar{k}/k)$ denote the Galois group of $\bar{k}$ over k with the Frobenius generator Φ. For any Γ-module M, let M^Γ (resp. M_Γ) denote the invariant (resp. coinvariant) subspace of M under Γ, that is, the kernel (resp. cokernel) of $\Phi - 1 : M \to M$. For any Abelian group M, M_{tors} denotes the torsion subgroup of M.

First we consider prime ℓ such that $(\ell, mp) = 1$. The results of Milne [Mil75, Mil86, Mil88], Schneider [Sc82] (cf. Bayer and Neukirch [BN78]) and Etesse [Et88] can be passed on to twisted Fermat motives as the cohomology group functors appearing in their formulas commute with the motivic decomposition.

PROPOSITION 6.7

Let ℓ be a prime such that $(\ell, mp) = 1$. Let $\mathcal{V}_A$ be a twisted Fermat motive of dimension n and degree m with twist $\mathbf{c}$ over $k = \mathbb{F}_q$.

1. Assume that $\mathcal{V}_A$ is supersingular but not strongly supersingular. Then for each integer r, $0 \le r \le n$,

$$\left| \mathrm{Norm}_{L/\mathbb{Q}} \left(1 - \frac{\vartheta(\mathbf{c}, \mathbf{a})}{q^r} \right) \right|_\ell^{-1} = 1.$$

2. Assume that m is a prime and that $\mathcal{V}_A$ is not supersingular. Then for each r, $0 \le r \le n$, the following assertions hold:

   (a) If $n = 2d$, then

   $$\left| B^r(\mathcal{V}_A) \right|_\ell^{-1} = \begin{cases} \#H^n(\mathcal{V}_{A_{\bar{k}}}, \mathbb{Z}_\ell(r))_\Gamma & \text{if } r \ne d, \\ \#H^n(\mathcal{V}_{A_{\bar{k}}}, \mathbb{Z}_\ell(r))_{\Gamma,\mathrm{tors}} & \text{if } r = d. \end{cases}$$

   (b) If $n = 2d + 1$, then

   $$\left| D^r(\mathcal{V}_A) \right|_\ell^{-1} = \#H^n(\mathcal{V}_{A_{\bar{k}}}, \mathbb{Z}_\ell(r))_\Gamma.$$

   In both cases, all the quantities appearing in the formulas are finite, and all cohomology groups are with respect to étale topology.

The main idea in the proof of the above proposition is to observe that the results of Milne [Mil86], §6, Milne [Mil88], §6 and Schneider [Sc82] (cf. Bayer and Neukirch [BN78]) can be passed on to motives. We state those results, specialized to our case, as a lemma.

LEMMA 6.8

Let ℓ be a prime such that $(\ell, mp) = 1$. Then for any integer r, $0 \le r \le n$, we have the following formulas:

1. Suppose that $2r \ne n$. Then

$$\left| Q(\mathcal{V}_A, q^{-r}) \right|_\ell^{-1} = \frac{\#H^n(\mathcal{V}_{A_{\bar{k}}}, \mathbb{Z}_\ell(r))_\Gamma}{\#H^n(\mathcal{V}_{A_{\bar{k}}}, \mathbb{Z}_\ell(r))^\Gamma}$$

where all the cohomology groups are finite.

2. Suppose that $2r = n$. Then

$$\left| Q(\mathcal{V}_A, q^{-r}) \right|_\ell^{-1} = \frac{\#H^n(\mathcal{V}_{A_{\bar{k}}}, \mathbb{Z}_\ell(r))_{\Gamma,\mathrm{tors}}}{[\#H^n(\mathcal{V}_{A_{\bar{k}}}, \mathbb{Z}_\ell(r))_{\mathrm{tors}}^\Gamma]^2}$$

where all the quantities are finite.

Now we use this to prove the proposition.

Proof of Proposition 6.7:

1. This is clear from Lemma 6.1.

2. By Theorem 6.2, we have, for each r, $0 \le r \le n$,

$$|Q(\mathcal{V}_A, q^{-r})|_\ell^{-1} = \begin{cases} |B^r(\mathcal{V}_A)|_\ell^{-1} & \text{if } n = 2d \\ |D^r(\mathcal{V}_A)|_\ell^{-1} & \text{if } n = 2d+1. \end{cases}$$

Now by Lemma 3.5, $\#H^i(\mathcal{V}_A, \mathbb{Z}_\ell(r)) = 1$ for any $i \ne n, n+1$. Moreover, there are isomorphisms:

$$H^n(\mathcal{V}_A, \mathbb{Z}_\ell(r)) \longrightarrow H^n(\mathcal{V}_{A_{\bar{k}}}, \mathbb{Z}_\ell(r))^\Gamma$$

and

$$H^{n+1}(\mathcal{V}_A, \mathbb{Z}_\ell(r)) \longleftarrow H^n(\mathcal{V}_{A_{\bar{k}}}, \mathbb{Z}_\ell(r))_\Gamma.$$

Observe also that ϕ acts semi-simply on $H^n(\mathcal{V}_{A_{\bar{k}}}, \mathbb{Q}_\ell(r))$, $0 \le r \le n$.

   (a) Let $n = 2d$. If $r \ne d$, then $H^{n+1}(\mathcal{V}_A, \mathbb{Z}_\ell(r)) = H^n(\mathcal{V}_{A_{\bar{k}}}, \mathbb{Z}_\ell(r))_\Gamma$ is finite, and if $r = d$, then $H^{n+1}(\mathcal{V}_A, \mathbb{Z}_\ell(r))_{\text{tors}}$ is finite. Furthermore, $H^n(\mathcal{V}_{A_{\bar{k}}}, \mathbb{Z}_\ell(r))$ is torsion-free so that $\#H^n(\mathcal{V}_A, \mathbb{Z}_\ell(r)) = \#H^n(\mathcal{V}_{A_{\bar{k}}}, \mathbb{Z}_\ell(r))^\Gamma = 1$. These facts together with Lemma 6.8 yield the required formulas.

   (b) Let $n = 2d + 1$. Then the assertion follows from the same line of argument as in the previous case with $r \ne d$. $\square$

Now we consider the prime $p = \text{char}(k)$.

PROPOSITION 6.9
Let $\mathcal{V}_A$ be a twisted Fermat motive of dimension n and degree m with twist $\mathbf{c}$ over $k = \mathbb{F}_q$.

1. *Assume that $\mathcal{V}_A$ is supersingular but not strongly supersingular. Then, for any r, $0 \le r \le n$,*

$$\left| \text{Norm}_{L/\mathbb{Q}}(1 - \frac{\mathfrak{J}(\mathbf{c}, \mathbf{a})}{q^r}) \right|_p^{-1} = 1.$$

2. *Assume that $\mathcal{V}_A$ is not supersingular but of Hodge–Witt type. Then for any r, $0 \le r \le n$, we have the following assertions:*

   (a) *Let $n = 2d$. Then*

$$|B^r(\mathcal{V}_A)|_p^{-1} = \begin{cases} \#H^n(\mathcal{V}_{A_{\bar{k}}}, \mathbb{Z}_p(r))_\Gamma & \text{if } r \ne d \\ \#H^n(\mathcal{V}_{A_{\bar{k}}}, \mathbb{Z}_p(r))_{\Gamma,\text{tors}} & \text{if } r = d. \end{cases}$$

(b) *Let* $n = 2d + 1$. *Then for each* r, $0 \leq r \leq n$,

$$|D^r(\mathcal{V}_A)|_p^{-1} = \#H^n(\mathcal{V}_{A_{\bar{k}}}, \mathbb{Z}_p(r))_\Gamma.$$

Once again, we begin by noting that the results of Milne [Mil86], §6 (cf. Etesse [Et88], Suwa and Yui [SY88], and Suwa [Su91b]) can be passed on to motives, and state them as a lemma.

LEMMA 6.10
For any r, $0 \leq r \leq n$, *we have the following formulas:*

1. *Suppose that* $2r \neq n$. *Then*

$$|q^{w(r)}Q(\mathcal{V}_A, q^{-r})|_p^{-1} = \frac{\#H^n(\mathcal{V}_{A_{\bar{k}}}, \mathbb{Z}_p(r))_\Gamma}{\#H^n(\mathcal{V}_{A_{\bar{k}}}, \mathbb{Z}_p(r))^\Gamma} + |q|_p^{-1}T^{r-1,n-r+1}(\mathcal{V}_A).$$

where all the quantities are finite.

2. *Suppose that* $2r = n$. *Then*

$$|q^{w(r)}Q(\mathcal{V}_A, q^{-r})|_p^{-1} = \frac{\#H^n(\mathcal{V}_{A_{\bar{k}}}, \mathbb{Z}_p(r))_{\Gamma,\text{tors}}}{[\#H^n(\mathcal{V}_{A_{\bar{k}}}, \mathbb{Z}_p(r))_{\text{tors}}^\Gamma]^2} + |q|_p^{-1}T^{r-1,n-r+1}(\mathcal{V}_A)$$

where all the quantities are finite.

In both cases, if $\mathcal{V}_A$ *is of Hodge–Witt type, then* $T^{r-1,n-r+1}(\mathcal{V}_A) = 0$.

Proof: (1) If $2r \neq n$, the formula of Milne [Mil86], Proposition (6.4) is read for a twisted Fermat motives $\mathcal{V}_A$ as follows:

$$Q(\mathcal{V}_A, q^{-r}) = \pm \frac{\#H^n(\mathcal{V}_{A_{\bar{k}}}, \mathbb{Z}_p(r))_\Gamma}{\#H^n(\mathcal{V}_{A_{\bar{k}}}, \mathbb{Z}_p(r))^\Gamma} \cdot q^{e^n(r)}$$

where

$$e^n(r) = T^{r,n-r}(\mathcal{V}_A) - \sum_{\substack{\mathbf{a} \in A \\ \nu(\mathcal{J}(\mathbf{c},\mathbf{a})) < r}} (r - \nu(\mathcal{J}(\mathbf{c},\mathbf{a})) = -w(r) + T^{r-1,n-r+1}(\mathcal{V}_A),$$

as $T^{r,n-r}(\mathcal{V}_A) = 0$. This gives the required formula.

(2) If $2r = n$, the formula of Milne [Mil88], Proposition (6.4) (see also Etesse [Et88]), is read as follows:

$$Q(\mathcal{V}_A, q^{-r}) = \frac{\#H^{n+1}(\mathcal{V}_A, \mathbb{Z}_p(r))_{\text{tors}}}{[\#H^n(\mathcal{V}_{A_{\bar{k}}}, \mathbb{Z}_p(r))_{\text{tors}}^\Gamma]^2} q^{-\alpha_r(\mathcal{V}_A)}$$

where

$$\alpha_r(\mathcal{V}_A) = T^{r-1,n-r+1}(\mathcal{V}_A) - T^{r,n-r}(\mathcal{V}_A) + \sum_{\substack{\mathbf{a} \in A \\ \nu(\mathcal{J}(\mathbf{c},\mathbf{a})) < r}} (r - \nu(\mathcal{J}(\mathbf{c},\mathbf{a}))$$

$$= T^{r-1,n-r+1}(\mathcal{V}_A) + w(r) - T^{r-1,n-r+1}(\mathcal{V}_A) = w(r).$$

This gives the required formula. □

Proof of Proposition 6.9: The first statement is clear from Lemma 6.1. For the second, note that we have, for each r, $0 \le r \le n$,

$$|Q(\mathcal{V}_A, q^{-r})|_p^{-1}$$

$$= \Big| \prod_{\substack{\mathbf{a} \in A \\ \nu(\Im(\mathbf{c},\mathbf{a}))=r}} (1 - \frac{\Im(\mathbf{c},\mathbf{a})}{q^r}) \Big|_p^{-1} \Big| \prod_{\substack{\mathbf{a} \in A \\ \nu(\Im(\mathbf{c},\mathbf{a}))<r}} (1 - \frac{\Im(\mathbf{c},\mathbf{a})}{q^r}) \Big|_p^{-1} \Big| \prod_{\substack{\mathbf{a} \in A \\ \nu(\Im(\mathbf{c},\mathbf{a}))>r}} (1 - \frac{\Im(\mathbf{c},\mathbf{a})}{q^r}) \Big|_p^{-1}.$$

Observe that

$$\Big| 1 - \frac{\Im(\mathbf{c},\mathbf{a})}{q^r} \Big|_p^{-1} = \begin{cases} \big| \frac{\Im(\mathbf{c},\mathbf{a})}{q^r} \big|_p^{-1} & \text{if} \quad \nu(\Im(\mathbf{c},\mathbf{a})) < r \\ 1 & \text{if} \quad \nu(\Im(\mathbf{c},\mathbf{a})) > r. \end{cases}$$

Then using the identity in Remark 6.3, we have

$$|q^{w(r)} Q(\mathcal{V}_A, q^{-r})|_p^{-1} = |q|_p^{-1} T^{r-1,n-r+1}(\mathcal{V}_A) + \Big| \prod_{\substack{\mathbf{a} \in A \\ \nu(\Im(\mathbf{c},\mathbf{a}))=r}} (1 - \frac{\Im(\mathbf{c},\mathbf{a})}{q^r}) \Big|_p^{-1}.$$

So we have only to interpret the term $\Big| \prod_{\substack{\mathbf{a} \in A \\ \nu(\Im(\mathbf{c},\mathbf{a}))=r}} (1 - \frac{\Im(\mathbf{c},\mathbf{a})}{q^r}) \Big|_p^{-1}$ in terms of some cohomological quantities.

Now by the first assertion in Lemma 3.5, $H^i(\mathcal{V}_{A_{\bar{k}}}, \mathbb{Z}_p(r)) = 0$ except for $i = n$ and $n + 1$. Moreover, there is an isomorphism

$$H^n(\mathcal{V}_A, \mathbb{Z}_p(r)) \longrightarrow H^n(\mathcal{V}_{A_{\bar{k}}}, \mathbb{Z}_p(r))^\Gamma$$

and also there is an exact sequence

$$0 \longrightarrow H^n(\mathcal{V}_{A_{\bar{k}}}, \mathbb{Z}_p(r))_\Gamma \longrightarrow H^{n+1}(\mathcal{V}_A, \mathbb{Z}_p(r)) \longrightarrow H^{n+1}(\mathcal{V}_{A_{\bar{k}}}, \mathbb{Z}_p(r))^\Gamma \longrightarrow 0.$$

1. Let $n = 2d$. If $r \ne d$, then

$$\Big| \prod_{\substack{\mathbf{a} \in A \\ \nu(\Im(\mathbf{c},\mathbf{a}))=r}} (1 - \frac{\Im(\mathbf{c},\mathbf{a})}{q^r}) \Big|_p^{-1} = \#H^n(\mathcal{V}_{A_{\bar{k}}}, \mathbb{Z}_p(r))_\Gamma.$$

Now, $H^{n+1}(\mathcal{V}_A, \mathbb{Z}_p(r))$ is finite, and we know that $H^{n+1}(\mathcal{V}_{A_{\bar{k}}}, \mathbb{Z}_p(r))^\Gamma$ has order $q^{T^{r-1,n-r+1}(\mathcal{V}_A)}$. Hence, $H^n(\mathcal{V}_{A_{\bar{k}}}, \mathbb{Z}_p(r))_\Gamma$ is also finite. Further, $\#H^n(\mathcal{V}_{A_{\bar{k}}}, \mathbb{Z}_p(r))^\Gamma = 1$. These facts, together with the first formula in Lemma 6.10, yield the required formula.

If $r = d$, then we have

$$\Big| \prod_{\substack{\mathbf{a} \in A \\ \nu(\Im(\mathbf{c},\mathbf{a}))=r}} (1 - \frac{\Im(\mathbf{c},\mathbf{a})}{q^r}) \Big|_p^{-1} = \#H^n(\mathcal{V}_{A_{\bar{k}}}, \mathbb{Z}_p(r))_{\Gamma,\text{tors}},$$

and we know that $H^{n+1}(\mathcal{V}_{A_{\bar{k}}}, \mathbb{Z}_p(r))^{\Gamma}$ is finite, and in fact that the order of $H^{n+1}(\mathcal{V}_{A_{\bar{k}}}, \mathbb{Z}_p(r))^{\Gamma}$ is equal to $q^{T^{r-1,n-r+1}(\mathcal{V}_A)}$. It follows that the group $H^n(\mathcal{V}_{A_{\bar{k}}}, \mathbb{Z}_p(r))_{\Gamma, \text{tors}}$ is finite. Further, we have $\#H^n(\mathcal{V}_{A_{\bar{k}}}, \mathbb{Z}_p(r))^{\Gamma}_{\text{tors}} = 1$. Combined with the second formula in Lemma 6.10, this yields the required formula.

2. If $n = 2d + 1$, the same argument as for (1), $2r \neq n$, gives the formula in question. $\square$

We now apply this to the computation of the central special value and the number $B^d(\mathcal{V}_A)$.

THEOREM 6.11
Let $\mathcal{V}_A$ be a twisted Fermat motive of dimension $n = 2d$ and prime degree $m > 3$ with twist $\mathbf{c}$ over $k = \mathbb{F}_q$. Suppose that $\mathcal{V}_A$ is of Hodge–Witt type and not supersingular. Then $B^d(\mathcal{V}_A)$ is a square, up to powers of m.
If Conjecture 1.9 holds, or if $\mathbf{c^a} \notin (k^{\times})^m$ for $\mathbf{a} \in A$, then $B^d(\mathcal{V}_A)$ is a square.

Proof: Let ℓ be a prime such that $(\ell, mp) = 1$. Then the Poincaré duality of the ℓ-adic étale cohomology groups and Proposition 6.7, part (2a), implies that the ℓ-part $|B^d(\mathcal{V}_A)|_{\ell}^{-1}$ is even.

For the prime $p = \text{char}(k)$, the duality of the p-adic cohomology groups and Proposition 6.9, part (2a), imply that the p-part $|B^d(\mathcal{V}_A)|_p^{-1}$ is even. Thus, $B^d(\mathcal{V}_A)$ is a square up to powers of m, as claimed.

For the m-part, if Conjecture 1.9 on the m-adic order of the Jacobi sums is true, then $|B^d(\mathcal{V}_A)|_m^{-1}$ is also even. Furthermore, if $\mathbf{c^a} \notin (k^{\times})^m$ for some (and hence any) $\mathbf{a} \in A$, then $B^d(\mathcal{V}_A)$ is relatively prime to m by Proposition 1.10, so that $|B^d(\mathcal{V}_A)|_m^{-1} = 1$. Therefore, if either Conjecture 1.9 holds (in fact, it need only hold for the $\mathbf{a}$) or if $\mathbf{c^a}$ is not an m-th power, $B^d(\mathcal{V}_A)$ is a square. $\square$

In fact, when $\mathbf{c^a}$ is an m-th power, the claim that the m-adic valuation of $B^d(\mathcal{V}_A)$ is even is clearly equivalent to Conjecture 1.9 for that particular $\mathbf{a}$.

DEFINITION 6.12 *Let $\mathcal{V}_A$ be a twisted Fermat motive of dimension $n = 2d$ and degree $m > 3$ with twist $\mathbf{c}$ over $k = \mathbb{F}_q$. Suppose that $\mathcal{V}_A$ is of Hodge–Witt type but not supersingular. Then the number $B^d(\mathcal{V}_A)$ defined in Theorem 6.2 is called the* Brauer number *of $\mathcal{V}_A$.*

This terminology was suggested to us by B. Mazur. The Brauer number $B^d(\mathcal{V}_A)$ should be related to the order of the "Brauer group" $\text{Br}^d(\mathcal{V}_A)$. This would be defined as the projection via the idempotent attached to A of the Brauer group considered by Milne and Lichtenbaum. It is not known even that $\text{Br}^d(\mathcal{V})$ exists, though it is known that its order must be a square if it does exist. It is not clear whether the same holds for the Brauer group of the motive $\mathcal{V}_A$, but Theorem 6.11 and our computations certainly suggest that this is the case. The results announced by Suwa [Su] suggest that one may want to extend the definition to the non-Hodge–Witt case. We refer the reader to his paper.

EXAMPLE 6.13 We now list some sample computational results on the Brauer numbers $B^d(\mathcal{V}_A)$ defined above. Notice that in every case the number we compute is a square. More examples and some discussion of methods for computing $B^d(\mathcal{V}_A)$ can be found in Section A.4.

(I) Take $(m,n) = (5,8)$. For simplicity, we restrict ourselves to the case when $q = p$, so that we are working over the prime field. This restricts us to primes p such that $p \equiv 1 \pmod 5$, and also makes all our motives automatically ordinary. In the table below, we let $q = p \in \{11, 31, 41, 61, 71\}$.

(1) Let $\mathbf{a} = (1,1,1,1,1,1,1,1,1,1) \in \mathfrak{A}_8^5$. The "Brauer number" $B^4(\mathcal{V}_A)$ is defined by the formula

$$\text{Norm}(1 - \mathfrak{J}(\mathbf{c},\mathbf{a})/p^4) = \frac{B^4(\mathcal{V}_A) \cdot 5}{p^4}.$$

The following table lists the values of $B^4(\mathcal{V}_A)$ for various values of the twist $\mathbf{c}$ and various primes:

twist	$p = 11$	$p = 31$	$p = 41$	$p = 61$	$p = 71$
$\mathbf{c} = (1,1,1,1,1,1,1,1,1,1)$	5^4	5^6	$3^4 \cdot 5^4$	$5^4 \cdot 109^2$	5^4
$\mathbf{c} = (1,1,1,1,1,1,1,1,1,3)$	139^2	$2^4 \cdot 109^2$	$3^4 \cdot 5^4$	3181^2	3919^2
$\mathbf{c} = (1,1,1,1,1,1,1,1,3,3)$	3^4	$3^4 \cdot 11^4$	$3^4 \cdot 5^4$	$3^4 \cdot 461^4$	$3^4 \cdot 821^2$
$\mathbf{c} = (1,1,1,1,1,1,1,3,3,3)$	$2^4 \cdot 19^2$	1511^2	$3^4 \cdot 5^4$	$19^2 \cdot 239^2$	$31^2 \cdot 59^2$
$\mathbf{c} = (1,2,3,4,1,2,3,4,1,2)$	139^2	5^6	2411^2	2^{12}	$2^4 \cdot 11^6$
$\mathbf{c} = (1,1,2,2,3,3,4,4,4,1)$	$2^4 \cdot 19^2$	139^2	401^2	$5^4 \cdot 109^2$	5^4
$\mathbf{c} = (1,1,1,1,2,2,2,3,3,3)$	181^2	5^6	401^2	3181^2	3919^2
$\mathbf{c} = (1,3,3,3,3,2,2,2,2,4)$	139^2	1511^2	941^2	$5^4 \cdot 109^2$	5^4

Recall that Iwasawa's congruence implies that whenever the twist is trivial $B^4(\mathcal{V}_A)$ will be divisible by m^2; in this case, it is in fact divisible by m^4. This implies that we have $\text{ord}_{(1-\zeta)}(j(\mathbf{a}) - p^4) = 5$, which is odd as predicted by Conjecture 1.9, and therefore $\text{ord}_{(1-\zeta)}(1 - j(\mathbf{a})) = 4 = \text{ord}_{(1-\zeta)}(p^4 - 1)$ for each of the primes p in the table. This explains why Conjecture 1.9 has to consider $\text{ord}_{(1-\zeta)}(j(\mathbf{a}) - q^d)$ rather than the simpler $\text{ord}_{(1-\zeta)}(j(\mathbf{a}) - 1)$.

We have checked also that if $p = 101$ we have $\text{ord}_{(1-\zeta)}(1 - j(\mathbf{a})) = 5$ (and, of course, $\text{ord}_{(1-\zeta)}(p^4 - 1) = 8$), so that once again we have $\text{ord}_{(1-\zeta)}(j(\mathbf{a}) - p^4) = 5$, once again confirming Conjecture 1.9.

The reader will note that 3 is a fifth power modulo 41, which explains several of the entries in that column.

(2) Now consider a different character $\mathbf{a} = (1,1,1,1,1,1,1,2,3,3) \in \mathfrak{A}_8^5$. Then $B^4(\mathcal{V}_A)$ is defined by

$$\text{Norm}(1 - \mathfrak{J}(\mathbf{c},\mathbf{a})/p^4) = \frac{B^4(\mathcal{V}_A) \cdot 5}{p^3},$$

as before. The following table lists the Brauer numbers $B^4(\mathcal{V}_A)$ for the same twists $\mathbf{c}$ and primes p considered above.

twist	$p = 11$	$p = 31$	$p = 41$	$p = 61$	$p = 71$
$\mathbf{c} = (1,1,1,1,1,1,1,1,1,1)$	$2^4 \cdot 5^2$	$2^4 \cdot 3^4 \cdot 5^2$	$2^4 \cdot 19^2 \cdot 5^2$	$2^4 \cdot 3^4 \cdot 5^2$	$2^4 \cdot 5^2$
$\mathbf{c} = (1,1,1,1,1,1,1,1,1,3)$	41^2	79^2	$2^4 \cdot 19^2 \cdot 5^2$	11^4	$7^4 \cdot 11^2$
$\mathbf{c} = (1,1,1,1,1,1,1,1,3,3)$	1	19^2	$2^4 \cdot 19^2 \cdot 5^2$	691^2	281^2
$\mathbf{c} = (1,1,1,1,1,1,1,3,3,3)$	41^2	79^2	$2^4 \cdot 19^2 \cdot 5^2$	11^4	$7^4 \cdot 11^2$
$\mathbf{c} = (1,2,3,4,1,2,3,4,1,2)$	29^2	19^2	79^2	11^4	$7^4 \cdot 11^2$
$\mathbf{c} = (1,1,2,2,3,3,4,4,4,1)$	$2^4 \cdot 5^2$	79^2	$11^2 \cdot 31^2$	691^2	281^2
$\mathbf{c} = (1,1,1,1,2,2,2,3,3,3)$	29^2	$2^4 \cdot 3^4 \cdot 5^2$	89^2	691^2	281^2
$\mathbf{c} = (1,3,3,3,3,2,2,2,2,4)$	$2^4 \cdot 5^2$	79^2	89^2	601^2	$11^2 \cdot 19^2$

(II) Let $(m,n) = (7,6)$ and let $q = p \in \{29, 43, 71, 113\}$. Note that $p \equiv 1$ (mod 7) in all cases, which implies that all our motives are ordinary.

(1) Let $\mathbf{a} = (1,1,1,1,1,1,4,4) \in \mathfrak{A}_6^7$. Again,

$$\text{Norm}(1 - \mathfrak{J}(\mathbf{c},\mathbf{a})/p^3) = \frac{B^3(\mathcal{V}_A) \cdot 7}{p^4}.$$

The following table lists the Brauer numbers $B^3(\mathcal{V}_A)$ for various values of the twist $\mathbf{c}$ and various $p = 29, 43, 71,$ and 113.

twist	$p = 29$	$p = 43$	$p = 71$	$p = 113$
$\mathbf{c} = (1,1,1,1,1,1,1,1)$	$97^2 \cdot 7^2$	$29^2 \cdot 7^2$	$41^2 \cdot 7^2$	$13^2 \cdot 167^2 \cdot 7^2$
$\mathbf{c} = (1,1,1,1,1,1,1,3)$	13^4	$2^{12} \cdot 13^2$	$41^2 \cdot 83^2$	$41^2 \cdot 83^2$
$\mathbf{c} = (1,1,1,1,1,1,3,3)$	$3^6 \cdot 41^2$	$3^6 \cdot 13^2$	$3^6 \cdot 13^2 \cdot 29^2$	$2^{12} \cdot 3^6$
$\mathbf{c} = (1,1,1,1,1,3,3,3)$	$2^{12} \cdot 13^2$	$29^2 \cdot 97^2$	$43^2 \cdot 239^2$	811^2
$\mathbf{c} = (1,2,3,4,5,6,1,2)$	$3^6 \cdot 41^2$	239^2	$13^2 \cdot 337^2$	$13^2 \cdot 29^2 \cdot 71^2$
$\mathbf{c} = (1,2,3,3,4,4,5,5)$	181^2	$3^6 \cdot 13^4$	$3^6 \cdot 13^2 \cdot 29^2$	$71^2 \cdot 139^2$
$\mathbf{c} = (1,1,1,2,2,3,3,3)$	$3^6 \cdot 41^2$	$29^2 \cdot 7^2$	$13^2 \cdot 181^2$	$71^2 \cdot 139^2$
$\mathbf{c} = (1,4,4,4,4,1,1,1)$	43^2	$29^2 \cdot 71^2$	$41^2 \cdot 83^2$	$71^2 \cdot 139^2$

(2) Let $\mathbf{a} = (1,1,1,2,3,3,5,5) \in \mathfrak{A}_6^7$. Write

$$\text{Norm}(1 - \mathcal{J}(\mathbf{c}, \mathbf{a})/p^3) = \frac{B^3(\mathcal{V}_A) \cdot 7}{p^2}.$$

The following table lists the values of the Brauer numbers $B^3(\mathcal{V}_A)$ for various values of the twist $\mathbf{c}$:

twist	$p = 29$	$p = 43$	$p = 71$	$p = 113$
$\mathbf{c} = (1,1,1,1,1,1,1,1)$	7^4	7^4	7^4	7^4
$\mathbf{c} = (1,1,1,1,1,1,1,3)$	13^2	83^2	71^2	97^2
$\mathbf{c} = (1,1,1,1,1,1,3,3)$	1	13^2	$2^6 \cdot 13^2$	43^2
$\mathbf{c} = (1,1,1,1,1,3,3,3)$	1	29^2	1	1
$\mathbf{c} = (1,2,3,4,5,6,1,2)$	1	13^2	7^4	43^2
$\mathbf{c} = (1,2,3,3,4,4,5,5)$	3^6	3^6	1	$2^6 \cdot 3^6$
$\mathbf{c} = (1,1,1,2,2,3,3,3)$	7^4	2^6	71^2	43^2
$\mathbf{c} = (1,4,4,4,4,1,1,1)$	7^4	7^4	7^4	7^4

REMARK 6.14 N. Boston pointed out to us that all prime factors, with the exception of small primes and of p, appearing in the Brauer numbers $B^d(\mathcal{V}_A)$ are of the form $\pm 1 \pmod{m}$. There is an elementary explanation of this fact.

Let ℓ be a prime such that $(\ell, m) = 1$. Let λ be a prime in L lying above ℓ. If ℓ divides the norm $\text{Norm}_{L/\mathbb{Q}}(1 - \mathcal{J}(\mathbf{c}, \mathbf{a})/q^d)$, then some conjugate of λ divides $1 - \mathcal{J}(\mathbf{c}, \mathbf{a})/q^d$. This implies that $\text{Norm}(\lambda)$ divides $\text{Norm}_{L/\mathbb{Q}}(1 - \mathcal{J}(\mathbf{c}, \mathbf{a})/q^d)$. But

Norm$(\lambda) = \ell^{\mathfrak{f}}$ where $\mathfrak{f}$ is the order of ℓ modulo m. So if ℓ^2 exactly divides the norm, then $\mathfrak{f} = 1$ or $\mathfrak{f} = 2$. In these cases, ℓ is of the form $\pm 1 \pmod m$. (However if ℓ^a divides the norm with $a > 2$, then ℓ is not necessarily of the form $\pm 1 \pmod m$.) Since our numbers are relatively small, the exponent of ℓ will tend to be 2 except for small primes, and that is what we see in the table.

We now compare the Brauer numbers associated to twisted Fermat motives $\mathcal{V}_A$ and those associated to Fermat motives $\mathcal{M}_A$ of the dimension n and degree m belonging to the $(\mathbb{Z}/m\mathbb{Z})^\times$-orbit A.

PROPOSITION 6.15
Let $\mathcal{V}_A$ (resp. $\mathcal{M}_A$) be a twisted Fermat motive of dimension n and degree m with twist $\mathbf{c}$ (resp. 1) over $k = \mathbb{F}_q$, belonging to the same $(\mathbb{Z}/m\mathbb{Z})^\times$-orbit A. Then for any r, $0 \le r \le n$, the following assertions hold:

1. $\text{Norm}_{L/\mathbb{Q}}(1 - \mathfrak{J}(\mathbf{c}, \mathbf{a})/q^r) = \text{Norm}_{L/\mathbb{Q}}(1 - j(\mathbf{a})/q^r)$ *if and only if we have* $\mathbf{c}^{\mathbf{a}} = c_0^{a_0}c_1^{a_1}\ldots c_{n+1}^{a_{n+1}} \in (k^\times)^m$ *for all* $\mathbf{a} \in A$.

2. *Let m be prime > 3, and $n = 2d$. If Conjecture 1.9 is true, then the quotient of the norms*

$$\frac{\text{Norm}_{L/\mathbb{Q}}(1 - j(\mathbf{a})/q^d)}{\text{Norm}_{L/\mathbb{Q}}(1 - \mathfrak{J}(\mathbf{c}, \mathbf{a})/q^d)}$$

is of the form m^{2e} with $e \ge 0$.

If $\mathbf{c}$ satisfies the property that $\mathbf{c}^{\mathbf{a}} \notin (k^\times)^m$ for $\mathbf{a} \in A$, then the assertion is true unconditionally.

Proof: The first assertion follows immediately from the definition of $\mathfrak{J}(\mathbf{c}, \mathbf{a})$. For the second, note first that both norms have the same p-adic order, and then that the m-part follows from Proposition 1.7, Theorem 6.11 and Remark 6.3, part (1). $\square$

EXAMPLE 6.16
1. Consider the same $\mathbf{c}, \mathbf{a}$ and p as in Example 6.6 on page 65. Then

$$\frac{\text{Norm}_{L/\mathbb{Q}}(1 - j(\mathbf{a})/p^2)}{\text{Norm}_{L/\mathbb{Q}}(1 - \mathfrak{J}(\mathbf{c}, \mathbf{a})/p^2)} = \begin{cases} 5^2 & \text{if } p = 11, 31 \\ 1 & \text{if } p = 41. \end{cases}$$

For $p = 41$, observe that $\mathbf{c}^{\mathbf{a}} \in (\mathbb{F}_{41}^\times)^5$.

2. Consider the same $\mathbf{c}$, $\mathbf{a}$ and p as in Example 6.13 on page 71, part (II). Observe that the m-part of the fraction

$$\frac{\text{Norm}_{L/\mathbb{Q}}(1 - j(\mathbf{a})/p^3)}{\text{Norm}_{L/\mathbb{Q}}(1 - \mathfrak{J}(\mathbf{c}, \mathbf{a})/p^3)}$$

is of the form m^{2e} with $e \ge 0$ for any twist $\mathbf{c}$. (In fact, it equals either 7^2 or 7^4 in all cases where $\mathbf{c}^{\mathbf{a}}$ is not an m-th power.)

Finally we discuss the effect of the inductive structures on the norms.

PROPOSITION 6.17

1. Let $\mathbf{a} \in \mathfrak{A}_n^m$ be a character, choose a non-zero $a \in \mathbb{Z}/m\mathbb{Z}$, and let $\tilde{\mathbf{a}} = (a_0, a_1, \ldots, a_{n+1}, a, m - a) \in \mathfrak{A}_{n+2}^m$ be the induced character of type I obtained from $\mathbf{a}$ and a. Then for any r such that $0 \leq r \leq n$, we have

$$\mathrm{Norm}_{L/\mathbb{Q}}(1 - \frac{\mathfrak{J}(\tilde{\mathbf{c}}, \tilde{\mathbf{a}})}{q^{r+1}}) = \mathrm{Norm}_{L/\mathbb{Q}}(1 - \frac{\mathfrak{J}(\mathbf{c}, \mathbf{a})}{q^r})$$

if and only if

$$\chi(-\frac{c_{n+2}}{c_{n+3}})^a = 1 \qquad \text{for all} \qquad \tilde{\mathbf{a}} \in \tilde{A}.$$

2. Let $m = m_0^t$ be a prime power where m_0 is a prime and either $m_0 > 3$ and $t \geq 2$ or $m_0 = 3$ and $t > 3$. Let $\mathbf{a} = (a_0, a_1, \ldots, a_{n+1}) \in \mathfrak{A}_n^m$ be a character with $\gcd(a_0, a_1, \ldots, a_{n+1}) = m_0$. Let $m' = m_0^{t-1}$ and consider the character $\mathbf{a}' = (a_0/m_0, a_1/m_0, \ldots, a_{n+1}/m_0) \in \mathfrak{A}_n^{m'}$. Then

$$\mathrm{Norm}_{\mathbb{Q}(\zeta_m)/\mathbb{Q}}(1 - \frac{\mathfrak{J}(\mathbf{c}, \mathbf{a})}{q^r}) = \mathrm{Norm}_{\mathbb{Q}(\zeta_{m'})/\mathbb{Q}}(1 - \frac{\mathfrak{J}(\mathbf{c}, \mathbf{a}')}{q^r})$$

for any r, $0 \leq r \leq n$.

Proof: Both statements follow immediately from Proposition 1.6. $\square$

7 Evaluating $Q(\mathcal{V}, T)$ at $T = q^{-r}$

Let $\mathcal{V} = \mathcal{V}_n^m(\mathbf{c})$ be a diagonal hypersurface of dimension n and degree $m > 3$ with twist $\mathbf{c}$ over $k = \mathbb{F}_q$. We want to evaluate the polynomial $Q(\mathcal{V}, T)$ at the critical point $T = q^{-r}$.

First we consider the even-dimensional case. Let $n = 2d$, and write

$$Q^*(\mathcal{V}, T) = (1 - q^d T)^{\rho_d(\mathcal{V}_k)} \prod (1 - \mathcal{J}(\mathbf{c}, \mathbf{a})T)$$

where the product is taken over all twisted Jacobi sums $\mathcal{J}(\mathbf{c}, \mathbf{a})$ such that $\mathcal{J}(\mathbf{c}, \mathbf{a}) \neq q^d$. Using the motivic decomposition, we can also write $Q^*(\mathcal{V}, T)$ in the following form

$$Q^*(\mathcal{V}, T) = (1 - q^d T)^{\rho_d(\mathcal{V}_k)} \prod Q(\mathcal{V}_A, T)$$

where the second product is taken over all twisted Fermat motives $\mathcal{V}_A$ which are not strongly supersingular.

To do this, we first recall, or further set up, some relevant notation.

$$\mathfrak{B}_n^m = \{\mathbf{a} \in \mathfrak{A} \mid \mathcal{J}(\mathbf{c}, \mathbf{a})/q^d = 1\},$$
$$\overline{\mathfrak{B}}_n^m = \{\mathbf{a} \in \mathfrak{A} \mid \mathcal{J}(\mathbf{c}, \mathbf{a})/q^d \ \text{ is a root of unity in } L\},$$
$$\mathfrak{C}_n^m = \overline{\mathfrak{B}}_n^m - \mathfrak{B}_n^m,$$
$$\mathfrak{D}_n^m = \mathfrak{A}_n^m - \mathfrak{B}_n^m,$$
$$\mathcal{O}(\mathfrak{C}_n^m) = \text{the set of } (\mathbb{Z}/m\mathbb{Z})^\times\text{-orbits in } \mathfrak{C}_n^m, \quad \text{and}$$
$$\mathcal{O}(\mathfrak{D}_n^m) = \text{the set of } (\mathbb{Z}/m\mathbb{Z})^\times\text{-orbits in } \mathfrak{D}_n^m$$

If $\mathcal{V} = \mathcal{V}_n^m$ with $n = 2d$, we further put

$$\varepsilon_d(\mathcal{V}_k) = \#\mathcal{O}(\mathfrak{C}_n^m) \quad (= \frac{\overline{\rho}_d(\mathcal{V}) - \rho_d(\mathcal{V})}{m - 1} \quad \text{if } m \text{ is prime})$$
$$\lambda_d(\mathcal{V}_{\overline{k}}) = \#\mathcal{O}(\mathfrak{D}_n^m) \quad (= \frac{B_n(\mathcal{V}) - \overline{\rho}_d(\mathcal{V})}{m - 1} \quad \text{if } m \text{ is prime}) \quad \text{and}$$
$$\delta_d(\mathcal{V}_k) = \varepsilon_d(\mathcal{V}_k) + \lambda_d(\mathcal{V}_{\overline{k}}) \quad (= \frac{B_n(\mathcal{V}) - \rho_d(\mathcal{V}_k)}{m - 1} \quad \text{if } m \text{ is prime}).$$

Our first result deals with supersingular diagonal hypersurfaces.

PROPOSITION 7.1

Let $(n = 2d)$. Let $\mathcal{V} = \mathcal{V}_n^m(\mathbf{c})$ be a supersingular diagonal hypersurface of dimension $n = 2d$ and prime degree $m > 3$ with twist $\mathbf{c}$ over $k = \mathbb{F}_q$. Then for each r, $0 \le r \le d$,

$$\lim_{s \to r} \frac{Q^*(\mathcal{V}, q^{-s})}{(1 - q^{d-s})\rho_d(\mathcal{V}_k)} = \prod_{\substack{\mathcal{V}_A \\ A \in O(\mathfrak{C}_n^m)}} \prod_{t \in (\mathbb{Z}/m\mathbb{Z})^\times} (1 - q^{d-r}\xi^t).$$

Proof: This follows immediately from Lemma 6.1. $\square$

COROLLARY 7.2

With the same conditions as in Proposition 7.1, the following assertions hold for $r = d$:

1. *If $\mathcal{V}_k$ is strongly supersingular, then the limit is equal to 1.*

2. *If $\mathcal{V}_k$ is supersingular, but not strongly supersingular, then the limit is equal to $m^{\varepsilon_d(\mathcal{V}_k)}$.*

Now we put together the motivic information we obtained in the previous chapter to obtain results about the special values of the zeta-function.

THEOREM 7.3

Let $n = 2d$. Let $\mathcal{V} = \mathcal{V}_n^m(\mathbf{c})$ be a diagonal hypersurface of dimension $n = 2d$ and prime degree $m > 3$ with twist $\mathbf{c}$ over $k = \mathbb{F}_q$. Suppose that $\mathcal{V}_k$ is of Hodge–Witt type. Then for any integer r, $0 \le r \le d$, we have

$$\lim_{s \to r} \frac{Q^*(\mathcal{V}, q^{-s})}{(1 - q^{d-s})\rho_d(\mathcal{V}_k)} = \frac{B^r(\mathcal{V}_k) \cdot m^{\lambda_d(\mathcal{V}_k)}}{q^{w_\mathcal{V}(r)}} \cdot \prod_{t \in (\mathbb{Z}/m\mathbb{Z})^\times} (1 - q^{d-r}\xi^t)^{\varepsilon_d(\mathcal{V}_k)},$$

where λ_d and ε_d are defined above, and the other quantities in the formula are defined as follows:

$$\rho_d(\mathcal{V}_k) = 1 + \#\{\mathbf{a} \in \mathfrak{A}_n^m \mid \mathcal{J}(\mathbf{c}, \mathbf{a}) = q^d\} \quad (= 1 \text{ for } \mathbf{c} \text{ extreme twist}),$$

$$\bar{\rho}_d(\mathcal{V}_k) = 1 + \sum B_n(\mathcal{V}_A)$$

where the sum is taken over all supersingular twisted Fermat motives $\mathcal{V}_A$, and

$$w_\mathcal{V}(r) = \sum_{i=0}^r (r - i) h^{i, n-i}(\mathcal{V}).$$

Here $B^r(\mathcal{V}_k)$ is a positive integer (not necessarily prime to mp) satisfying

$$B^r(\mathcal{V}_k) = B^{n-r}(\mathcal{V}_k).$$

Proof: For each twisted Fermat motive $\mathcal{V}_A$ which is not supersingular, we have

$$Q(\mathcal{V}_A, q^{-r}) = \operatorname{Norm}_{L/\mathbb{Q}}(1 - \frac{\mathfrak{J}(\mathbf{c}, \mathbf{a})}{q^r}) \in \frac{1}{q^{w(r)}}\mathbb{Z}$$

with

$$w(r) \, (= w_{\mathcal{V}_A}(r)) = rh^{0,n}(\mathcal{V}_A) + (r-1)h^{1,n-1}(\mathcal{V}_A) + \cdots + h^{r-1,n-r+1}(\mathcal{V}_A)$$

by Theorem 6.2. By Lemma 3.5 (2), the functor $H^j(\quad, \Omega^i)$ with $i + j = n$ commutes with the motivic decomposition $\tilde{\mathcal{V}} = \oplus \mathcal{V}_A$. Therefore, gluing the results of Theorem 6.2, the exponent of q in $Q^*(\mathcal{V}, q^{-r})$ is given by

$$\sum_{\substack{\mathcal{V}_A \\ \text{not s.s.}}} \sum_{i=0}^{r}(r-i)h^{i,n-i}(\mathcal{V}_A) = \sum_{i=0}^{r}(r-i)h^{i,n-i}(\mathcal{V})$$

where the first sum in the left-hand side runs over all twisted Fermat motives $\mathcal{V}_A$ which are not supersingular.

If m is prime > 3, then Theorem 6.2 (1) and the hypothesis that $q \equiv 1$ (mod m) yield the congruence

$$Q(\mathcal{V}_A, q^{-r}) \equiv 0 \pmod{m} .$$

Twisted Fermat motives $\mathcal{V}_A$ which are supersingular but not strongly supersingular give rise to the auxiliary factor

$$\left(\prod_{t \in (\mathbb{Z}/m\mathbb{Z})^\times} (1 - q^{d-r}\xi^t)\right)^{\varepsilon_d(\mathcal{V}_k)} \qquad (= m^{\varepsilon_d(\mathcal{V}_k)} \text{ if } r = d).$$

There are altogether $\lambda_d(\mathcal{V}_{\bar{k}})$ twisted Fermat motives $\mathcal{V}_A$ which are not supersingular. Thus the assertion on the m-part follows.

The assertion for $B^r(\mathcal{V}_k)$ follows from Proposition 6.4 and Corollary 6.5, noting that $B^r(\mathcal{V}_k) = \prod B^r(\mathcal{V}_A)$ where the product is taken over all twisted Fermat motives which are not strongly supersingular. $\square$

COROLLARY 7.4
With the same conditions as in Theorem 7.3, the following assertion holds: if $\mathcal{V}_k$ is of Hodge–Witt type, then for $r = d$, the limit is equal to

$$B^d(\mathcal{V}_k) \cdot m^{\delta_d(\mathcal{V}_k)}/q^{w_\mathcal{V}(d)}.$$

THEOREM 7.5
Let $n = 2d$. Let $\mathcal{V} = \mathcal{V}_n^m(\mathbf{c})$ be a diagonal hypersurface of dimension $n = 2d$ and prime degree $m > 3$ with twist $\mathbf{c}$ over $k = \mathbb{F}_q$. Suppose that $\mathcal{V}_k$ is of Hodge–Witt type. If Conjecture 1.9 holds for each $\mathbf{a} \in \mathfrak{A}_n^m$ such that $\mathbf{c}^{\mathbf{a}}$ is an m-th power, then the integer $B^d(\mathcal{V}_k)$ is a square.

In particular, if $\mathbf{c}$ is extreme, then $B^d(\mathcal{V}_k)$ is a square.

Proof: We have for any prime ℓ (including $\ell = p$),

$$|B^d(V_k)|_\ell^{-1} = \sum |B^d(V_A)|_\ell^{-1}$$

where the sum is taken over all twisted Fermat motives which are not strongly supersingular. Then the assertion follows from Theorem 6.11. $\square$

For the Fermat hypersurfaces case, results similar to the previous ones were announced by Suwa [Su91b]. Suwa [Su] has also announced results removing the restriction to hypersurfaces of Hodge–Witt type.

Now we consider odd-dimensional diagonal hypersurfaces $V = V_n^m(\mathbf{c})$ over $k = \mathbb{F}_q$. Let $n = 2d + 1$. For each r, $0 \le r \le d$,

$$Q(V, q^{-r}) = \prod Q(V_A, q^{-r})$$

where the product is taken over all twisted Fermat motives V_A.

THEOREM 7.6
Let $n = 2d + 1$. Let m be a prime > 3. For any integer r, $0 \le r \le d$, let

$$D^r(V_k) = q^{w_V(r)} \cdot Q(V, q^{-r}).$$

Then $D^r(V_k)$ is a positive integer (not necessarily prime to mp) such that

$$D^r(V_k) = D^{n-r}(V_k)$$

and

$$w_V(r) = \sum_{i=0}^{r} (r - i) h^{i, n-i}(V).$$

Proof: Observe that

$$D^r(V_k) = \prod_{V_A} D^r(V_A) = \prod \mathrm{Norm}_{L/\mathbb{Q}} (1 - \frac{\mathcal{J}(\mathbf{c}, \mathbf{a})}{q^r})$$

where the product is taken over all the motives V_A. Then the assertion follows from Lemma 6.1 and Theorem 6.2 (2). The duality on $D^r(V_k)$ follows from Proposition 6.4 and Corollary 6.5. $\square$

One can obtain cohomological interpretations of the integers $B^r(V_k)$ and $D^r(V_k)$ for $0 \le r \le n$ from the results of Milne [Mil86, Mil88].

PROPOSITION 7.7
Let $V = V_n^m(\mathbf{c})$ be a diagonal hypersurface of dimension n and degree $m > 3$ with twist $\mathbf{c}$ over $k = \mathbb{F}_q$. Assume that V is of Hodge–Witt type. Then for any integer r, $0 \le r \le n$, the following assertions hold:

1. Let $n = 2d$. Then

$$B^r(\mathcal{V}_k) = \begin{cases} \pm \# H^n(\mathcal{V}_{\bar{k}}, \hat{\mathbb{Z}}(r))_\Gamma & \text{if } r \neq d \\ \pm \# H^n(\mathcal{V}_{\bar{k}}, \hat{\mathbb{Z}}(r))_{\Gamma, \text{tors}} & \text{if } r = d. \end{cases}$$

2. Let $n = 2d + 1$. Then

$$D^r(\mathcal{V}_k) = \pm \# H^n(\mathcal{V}_{\bar{k}}, \hat{\mathbb{Z}}(r))_\Gamma.$$

All the cohomology groups appearing in the formulas are finite.

Proof: This puts together Proposition 6.7 and Proposition 6.9. $\square$

Some computations of the global "Brauer numbers" can be found in Section A.5.

We now compare the asymptotic values of the partial zeta-functions of a diagonal hypersurface $\mathcal{V} = \mathcal{V}_n^m(\mathbf{c})$ and the Fermat variety $\mathcal{X} = \mathcal{V}_n^m(\mathbf{1})$ over $k = \mathbb{F}_q$. An interesting case is when $n = 2d$.

PROPOSITION 7.8
Let $\mathcal{V} = \mathcal{V}_n^m(\mathbf{c})$ and $\mathcal{X} = \mathcal{V}_n^m(\mathbf{1})$ be a diagonal and the Fermat hypersurfaces of dimension $n = 2d$ and prime degree $m > 3$ with twist $\mathbf{c}$ and $\mathbf{1}$, respectively, over $k = \mathbb{F}_q$. Then the quotient

$$\lim_{s \to d} \left[\frac{Q(\mathcal{V}, q^{-s})}{(1 - q^{d-s})^{\rho_d(\mathcal{V}_k)}} \bigg/ \frac{Q(\mathcal{X}, q^{-s})}{(1 - q^{d-s})^{\rho_d(\mathcal{X}_k)}} \right]$$

is equal to

1. $m^{\varepsilon_d(\mathcal{V}_k) - \varepsilon_d(\mathcal{X}_k)}$ *if both $\mathcal{V}$ and $\mathcal{X}$ are supersingular, and*

2. $\frac{B^d(\mathcal{V}_k)}{B^d(\mathcal{X}_k)} \cdot m^{\delta_d(\mathcal{V}_k) - 3\delta_d(\mathcal{X}_k)}$ *if both $\mathcal{V}$ and $\mathcal{X}$ are of Hodge–Witt type.*

8 The Lichtenbaum–Milne conjecture

Consider diagonal hypersurfaces $\mathcal{V} = \mathcal{V}_n^m(\mathbf{c})$ of even dimension $n = 2d \geq 2$ and degree m with twist $\mathbf{c}$ over $k = \mathbb{F}_q$. As a higher-dimensional analogue of the Artin–Tate formula, Milne [Mil86, Mil88] and Lichtenbaum [Li84, Li87, Li90] have formulated a conjectural formula on the special value of the partial zeta-function of $\mathcal{V}$ at $T = q^{-d}$. In this section, we compare our results with those predicted by their formula.

The Milne–Lichtenbaum conjecture concerns the residue of the zeta-function $Z(\mathcal{V}, T)$ (or rather of the partial zeta-function $Q(\mathcal{V}, T)/(1 - q^r T)^{\rho_r(\mathcal{V}_k)}$) at integral arguments $T = q^{-r}$ for $0 \leq r \leq n$. Particularly interesting is the case when $r = d$. In this case, Milne [Mil86, Mil88] has a formula for the limit $Q^*(\mathcal{V}, q^{-s})/(1 - q^{d-s})^{\rho_d(\mathcal{V}_k)}$ as s tends to d, which holds if we assume the validity of the Tate conjecture, the existence of certain complexes of étale sheaves $\mathbb{Z}(d)$, and the surjectivity of the cycle map. Such complexes are used to define some motivic cohomology groups, and the candidates for them have been defined by Lichtenbaum for $d \leq 2$. See [Li84, Li87, Li90]. The existence of such complexes for $d > 2$ is still unknown.

We begin by stating the formula of Lichtenbaum and Milne for the partial zeta-function of $\mathcal{V}_k$.

THEOREM 8.1 (THE LICHTENBAUM–MILNE FORMULA)
Let $\mathcal{V} = \mathcal{V}_n^m(\mathbf{c})$ be a diagonal hypersurface of dimension $n = 2d \geq 2$ and degree m with twist $\mathbf{c}$ over $k = \mathbb{F}_q$. Let $\mathrm{CH}^d(\mathcal{V}_k)$ denote the Chow group of algebraic cycles of codimension d on $\mathcal{V}$ defined over k modulo algebraic equivalence. Assume that

1. *there exists a complex $\mathbb{Z}(d)$ satisfying the axioms in Milne [Mil86, Mil88], and that*

2. *the cycle map $\mathrm{CH}^d(\mathcal{V}_k) \to H^n(\mathcal{V}_k, \mathbb{Z}(d))$ is surjective.*

Then, if the Tate conjecture holds for $\mathcal{V}_k$, we have

$$\lim_{s \to d} \frac{Q^*(\mathcal{V}, q^{-s})}{(1 - q^{d-s})^{\rho_d(\mathcal{V}_k)}} = \pm \frac{\#\,\mathrm{Br}^d(\mathcal{V}_k)|\det A^d(\mathcal{V}_k)|}{q^{\alpha_d(\mathcal{V})}[\#A^d(\mathcal{V}_k)_{tor}]^2}$$

where the quantities on the right-hand side are explained as follows:

- $\mathrm{Br}^d(\mathcal{V}_k) = H^{2d+1}(\mathcal{V}_k, \mathbb{Z}(d))$ denotes the "Brauer group" of $\mathcal{V}_k$,

- $A^d(\mathcal{V}_k) = \mathrm{Im}[\mathrm{CH}^d(\mathcal{V}_k) \to H^n(\mathcal{V}_{\bar{k}}, \hat{\mathbb{Z}}(d))]$ is the image of the Chow group $\mathrm{CH}^d(\mathcal{V}_k)$ in the cohomology $H^n(\mathcal{V}_{\bar{k}}, \hat{\mathbb{Z}}(d))$,

- $\{D_i\}$ is a $\mathbb{Z}$-basis for $A^d(\mathcal{V}_k)$ modulo torsion,

- $\det A^d(\mathcal{V}_k) = \det(D_i \cdot D_j)$ is the determinant of the intersection pairing on $A^d(\mathcal{V}_k)$,

- $A^d(\mathcal{V}_k)_{\mathrm{tor}}$ is the torsion subgroup of $A^d(\mathcal{V}_k)$, and

- $\alpha_d(\mathcal{V}) = s^{n+1}(d) - 2s^n(d) + \sum_{\nu(\mathfrak{J}(\mathbf{c},\mathbf{a})<d}(d - \nu(\mathfrak{J}(\mathbf{c},\mathbf{a}))$, where we define $s^r(d) := \dim \underline{H}^r(\mathcal{V}_k, \mathbb{Z}_p(d))$ (as a perfect group scheme).

REMARK 8.2 For $n = 2$, the Tate conjecture holds for $\mathcal{V} = \mathcal{V}_2^m$ over k, and this formula is indeed the Artin–Tate formula:

- $A^1(\mathcal{V}_k) = \mathrm{NS}(\mathcal{V}_k) =$ the Néron–Severi group of $\mathcal{V}$,

- $\mathrm{Br}^1(\mathcal{V}_k) = H^3(\mathcal{V}_k, \mathbb{Z}(1)) = H^2(\mathcal{V}_k, \mathbb{G}_m)$ is the cohomological Brauer group of $\mathcal{V}_k$ (which is isomorphic to the algebraic Brauer group of $\mathcal{V}_k$),

- $\det A^1(\mathcal{V}_k) = \mathrm{disc}\, \mathrm{NS}(\mathcal{V}_k)$, and

- $\#A^1(\mathcal{V}_k)_{\mathrm{tor}} = 1$, $\alpha_1(\mathcal{V}) = p_g(\mathcal{V})$.

(See Tate [Ta68], Milne [Mil86, Mil88].)

From the Artin–Tate formula, we can deduce the following assertions.

COROLLARY 8.3
Let $\mathcal{V} = \mathcal{V}_2^m(\mathbf{c})$ be a diagonal hypersurface of dimension $n = 2$ and degree m with twist $\mathbf{c}$ over $k = \mathbb{F}_q$. Then the following assertions hold:

1. If $\mathcal{V}_k$ is supersingular, then

$$\#\mathrm{Br}^1(\mathcal{V}_k)|\,\mathrm{disc}\,\mathrm{NS}(\mathcal{V}_k)| = q^{p_g(\mathcal{V})} m^{\epsilon_1(\mathcal{V}_k)}$$

where $\epsilon_1(\mathcal{V}_k)$ is the quantity defined in Chapter 7.

In particular, if $\mathcal{V}_k$ is strongly supersingular, then $\mathrm{Br}^1(\mathcal{V}_k)$ is a p-group, and $\mathrm{disc}\,\mathrm{NS}(\mathcal{V}_k)$ divides a power of p.

2. Assume that m is a prime > 3. If $\mathcal{V}_k$ is of Hodge–Witt type, then

$$\#\mathrm{Br}^1(\mathcal{V}_k)|\,\mathrm{disc}\,\mathrm{NS}(\mathcal{V}_k)| = B^1(\mathcal{V}_k) \cdot m^{\delta_1(\mathcal{V}_k)},$$

where $B^1(\mathcal{V}_k)$ is defined as in Theorem 7.3, and $\delta_1(\mathcal{V}_k) = (m - 3)^2$.

(Cf. Shioda [Sh87], Suwa and Yui [SY88, SY89].)

Some of the quantities in Theorem 8.1 can be computed for diagonal hypersurfaces.

PROPOSITION 8.4
Let $V = V_n^m(\mathbf{c})$ be a diagonal hypersurface of dimension $n = 2d$ and degree m with twist $\mathbf{c}$ over $k = \mathbb{F}_q$. Then the following assertions hold:

1. $A^d(V_k)$ is torsion-free.

2. If V_k is of Hodge-Witt type, then $s^{n+1}(d) = s^n(d) = 0$, so that we have $\alpha_d(V) = w_V(d)$.

3. If V_k is supersingular, then $w_V(d) = 0$ and $\alpha_d(V) = s^{n+1}(d) - 2s^n(d)$.

Proof: (1) Since V is a complete intersection, $A^d(V)$ is torsion-free by Deligne [De73].

(2) If V_k is of Hodge–Witt type, then we have $H^n(V_k, \mathbb{Z}_p(d)) = 0$ and $H^{n+1}(V_k, \mathbb{Z}_p(d))$ is finite. So the assertion follows from Theorem 6.2 and Theorem 7.3.

(3) If V_k is supersingular, then $w_V(d) = 0$ as the Newton polygon has the pure slope d. In this case, the formal groups $\Phi^\bullet(V)$ are all unipotent by Proposition 3.11. (As noted by Milne [Mil86], Remark 3.5, the actual computation of the invariants $s^{n+1}(d)$ and $s^n(d)$ seems very difficult. In fact, we need to determine the structure of the formal groups attached to V, especially, the number of copies of $\widehat{\mathbb{G}}_a$ occurring in the formal groups.) □

In the case of an extreme twist, we can also determine the contribution from the intersection pairing:

PROPOSITION 8.5
Let $V = V_n^m(\mathbf{c})$ be a diagonal hypersurface of dimension $n = 2d$, prime degree $m > 3$ with an extreme twist $\mathbf{c}$ defined over a finite field $k = \mathbb{F}_q$. Then $A^d(V_k)$ is generated over $\mathbb{Q}$ by the single class $[H]$ consisting of hyperplane sections on V_k of codimension d, and any hyperplane section $H \in [H]$ has the self-intersection number
$$(H, H) = m.$$

Proof: Since $\mathbf{c}$ is extreme, we know $\rho_d'(V_k) = \rho_d(V_k) = 1$, and hence $A^d(V_k)$ is generated by a hyperplane section. Let $\mathcal{H}$ be a hyperplane of dimension $d+1$ and let $H =: V \cap \mathcal{H}$. Then we can compute the self-intersection number (H, H) by taking another hyperplane $\mathcal{H}'$ of dimension $d+1$ and looking at the multiplicities of the intersection $V \cap \mathcal{H}$ and $V \cap \mathcal{H}'$, that is, of $V \cap \mathcal{H} \cap \mathcal{H}'$. Now $V \cap \mathcal{H}$ is a subvariety of degree m and dimension d in projective space so that its intersection with a "generic" hyperplane $\mathcal{H}'$ consists of exactly m points. Therefore $(H, H) = m$. □

Observe that the assertion of Proposition 8.5 remains valid for any field k, of any characteristic as long as $\mathcal{V}_k$ has an extreme twist.

COROLLARY 8.6
With the conditions of Proposition 8.5, we have

$$|\det A^d(\mathcal{V}_k)| = m.$$

The final quantity in Theorem 8.1 is the "Brauer group" $\mathrm{Br}^d(\mathcal{V}_k)$. This is not even known to exist unless the complex $\mathbb{Z}(d)$ does, so we do not know whether the number we can compute is indeed the order of the Brauer group. The duality properties of $\mathbb{Z}(d)$ do imply, however, that this order (when it is defined) must be a square, and this is what we exploit below.

We consider first the case when $\mathcal{V}$ is of Hodge–Witt type.

THEOREM 8.7
Let $\mathcal{V} = \mathcal{V}_n^m(\mathbf{c})$ be a diagonal hypersurface of dimension $n = 2d$ and prime degree $m > 3$ with twist $\mathbf{c}$ over $k = \mathbb{F}_q$. Suppose $\mathcal{V}_k$ is of Hodge–Witt type. Then the Lichtenbaum–Milne formula holds if and only if we have

$$\# \mathrm{Br}^d(\mathcal{V}_k)|\det A^d(\mathcal{V}_k)| = m^{\delta_d(\mathcal{V}_k)} \cdot B^d(\mathcal{V}_k),$$

where we write

$$B^d(\mathcal{V}_k) = \prod B^d(\mathcal{V}_A),$$

the product being taken over all the non-supersingular motives. The number $B^d(\mathcal{V}_k)$ is a square up to powers of m.

If, in addition, $\mathbf{c}$ is an extreme twist, then we have $|\det A^d(\mathcal{V}_k)| = m$, so that the Lichtenbaum–Milne formula holds if and only if

$$\# \mathrm{Br}^d(\mathcal{V}_k) = m^{\delta_d(\mathcal{V}_k)-1} \cdot B^d(\mathcal{V}_k)$$

The exponent $\delta_d(\mathcal{V}_k) - 1$ is even, and $B^d(\mathcal{V}_k)$ is a square.

Proof: This is just a matter of putting together all that we have already proved. To see that the exponent of m is even in the extreme case, recall that

$$\delta(\mathcal{V}_k) = \frac{B_n(\mathcal{V}) - \rho_d(\mathcal{V}_k)}{m - 1},$$

and that

$$B_n(\mathcal{V}) = \frac{(m-1)^{n+2} + (m-1)}{m} + 1.$$

Since $\mathcal{V}_k$ is extreme, $\rho_d(\mathcal{V}_k) = 1$, and a direct calculation shows that $\delta_k(\mathcal{V}_k) - 1$ is even. $\square$

Note that when $n = 2$ the Tate conjecture is known to hold and the complex $\mathbb{Z}(1)$ is known to exist, so that the formula in Theorem 8.7 holds unconditionally and the number we have computed is indeed the order of the Brauer group of $\mathcal{V}_k$. When $n = 4$ there are many cases where the Tate conjecture is known to be true by Proposition 5.17 (e.g., when m is prime or $m \leq 21$ and $p \equiv 1$ (mod m)), and the existence of the complex $\mathbb{Z}(2)$ was established by Lichtenbaum in [Li87, Li90]. In these cases, then, the validity of the formula in Theorem 8.7 is conditional only on the surjectivity of the cycle map.

We have computed the Brauer number $B^d(\mathcal{V}_k)$ for many different $\mathcal{V}$ (of prime degree). In all cases, it turns out to be a square (including the m-part). It is natural, then, to conjecture that this is always the case.

CONJECTURE 8.8 *Let* $\mathcal{V} = \mathcal{V}_n^m(\mathbf{c})$ *be a diagonal hypersurface of dimension* $n = 2d$ *and prime degree* $m > 3$ *with twist* $\mathbf{c}$ *over* $k = \mathbb{F}_q$. *Suppose* $\mathcal{V}_k$ *is of Hodge–Witt type. Then the global "Brauer number"* $B^d(\mathcal{V}_k)$ *is a square.*

Note that Conjecture 8.8 follows from Conjecture 1.9. One should also remark that in the product expression for $B^d(\mathcal{V}_k)$ many of the motivic Brauer numbers $B^d(\mathcal{V}_A)$ appear with even multiplicity, so that Conjecture 8.8 is weaker than the statement that all the $B^d(\mathcal{V}_A)$ are squares (which also follows from Conjecture 1.9). In our calculations, the motivic Brauer numbers always turn out to be squares.

It is natural to ask about the exponent $\delta_d(\mathcal{V}_k)$. In the non-extreme case, one can make no predictions, since one doesn't know the value of the determinant $\det A^d(\mathcal{V}_k)$. In fact, even in the simplest case, one sees both odd and even exponents. For example, take $m = 5$, $n = 2$, $p = 11$. In table (b1) in Section A.3, one sees that for various twists one gets $\rho_d(\mathcal{V}_k) = 5$ or 9, which gives $\delta_d(\mathcal{V}_k) = 12$ and 11, respectively. For the trivial twist, we get $\rho_d(\mathcal{X}_k) = 37$ and $\delta_d(\mathcal{X}_k) = 4$.

Now we go on to consider the case where $\mathcal{V}_k$ is supersingular. This is a little less satisfactory, because we are unable to get an explicit value for $\alpha_d(\mathcal{V})$. So our results are necessarily partial.

THEOREM 8.9
Let $\mathcal{V} = \mathcal{V}_n^m(\mathbf{c})$ *be a diagonal hypersurface of dimension* $n = 2d$ *and prime degree* $m > 3$ *with twist* $\mathbf{c}$ *over* $k = \mathbb{F}_q$. *Suppose* $\mathcal{V}_k$ *is supersingular. Then the Lichtenbaum–Milne formula holds for* $\mathcal{V}_k$ *if and only if we have*

$$\# \operatorname{Br}^d(\mathcal{V}_k) |\det A^d(\mathcal{V}_k)| = q^{\alpha_d(\mathcal{V})} m^{\delta_d(\mathcal{V}_k)}.$$

In particular:

1. *If* $\mathcal{V}_k$ *is strongly supersingular, then* $\delta(\mathcal{V}_k) = 0$ *and the Lichtenbaum–Milne formula holds if and only if we have*

$$\# \operatorname{Br}^d(\mathcal{V}_k) |\det A^d(\mathcal{V}_k)| = q^{\alpha_d(\mathcal{V})}.$$

In particular, $\mathrm{Br}^d(\mathcal{V}_k)$ must be a p-group.

2. If $\mathbf{c}$ is extreme, then we have $|\det A^d(\mathcal{V}_k)| = m$ and the Lichtenbaum–Milne formula holds if and only if we have

$$\# \mathrm{Br}^d(\mathcal{V}_k) = q^{\alpha_d(\mathcal{V})} m^{\delta_d(\mathcal{V}_k)-1}.$$

The m-part of this number is a square.

Proof: Clear from the above. □

Note that since the Tate conjecture is known to hold when $\mathcal{V}_k$ is supersingular, we know that the Lichtenbaum–Milne formula holds for $n = 2$. For $n = 4$, the formulas in the theorem are conditional only on the surjectivity of the cycle map, since Lichtenbaum [Li87, Li90] has constructed the required complex $\mathbb{Z}(2)$.

We describe a sample computation. Let $m = 7$, $n = 6$, and let $\mathcal{V} = \mathcal{V}_6^7(\mathbf{c})$ over $\mathbb{F}_{29}$, where we take $\mathbf{c}$ to be the extreme twist $\mathbf{c} = (2, 1, 1, 1, 1, 1, 1, 1)$. Since m is prime and $p \equiv 1 \pmod{m}$, the Tate conjecture holds. Then the global "Brauer number" of $\mathcal{V}$ is computed as follows. One produces a "minimal" list of characters, i.e., a list of characters such that their associated motives make up a set of representatives of the isomorphism classes of motives $\mathcal{V}_A$. One then computes the norm for each of our characters $\mathbf{a}$, and then it is simply a matter of putting the data together (taking multiplicities into account). The m-part can be computed directly, since we have

$$\delta_d(\mathcal{V}_k) = \frac{(m-1)^{n+2} + (m-1)}{m(m-1)} = 39991,$$

so that, as pointed out above, $\delta_d(\mathcal{V}_k) - 1$ is even.

As to the global "Brauer number", we can compute the motivic "Brauer numbers" and then put them all together. The full list of values can be found in Section A.5. Putting them all together with the correct multiplicities gives

$$\begin{aligned}
B^3(\mathcal{V}_k) = &\, 2^{152220} \cdot 3^{25200} \cdot 5^{2268} \cdot 13^{53056} \cdot 29^{3024} \cdot 41^{16576} \cdot 43^{14392} \cdot \\
&\, 71^{1736} \cdot 83^{4144} \cdot 97^{1120} \cdot 223^{336} \cdot 281^{5320} \cdot 349^{168} \cdot 379^{280} \cdot \\
&\, 461^{840} \cdot 631^{56} \cdot 953^{336}.
\end{aligned}$$

Thus, the Lichtenbaum–Milne formula will hold if we have

$$\begin{aligned}
\# \mathrm{Br}^3(\mathcal{V}_k) = &\, 7^{39990} \cdot 2^{152220} \cdot 3^{25200} \cdot 5^{2268} \cdot 13^{53056} \cdot 29^{3024} \cdot 41^{16576} \cdot 43^{14392} \cdot \\
&\, 71^{1736} \cdot 83^{4144} \cdot 97^{1120} \cdot 223^{336} \cdot 281^{5320} \cdot 349^{168} \cdot 379^{280} \cdot \\
&\, 461^{840} \cdot 631^{56} \cdot 953^{336}.
\end{aligned}$$

It is interesting to remark that the multiplicities are even for every one of the motives in this case. This suggests that the fact that the Brauer number is a square is perhaps less interesting than the fact that the motivic Brauer numbers are squares. In fact, it suggests that there should exist a Brauer group of each motive $\mathcal{V}_A$, and that its order should be a square.

More examples can be found in Section A.5.

Now we compare the above results with the Lichtenbaum–Milne conjecture (Conjecture 0.2) on the residue of $Z(\mathcal{V}, q^{-s})$ as $s \to d$.

THEOREM 8.10
Let $\mathcal{V} = \mathcal{V}_n^m(\mathbf{c})$ be a diagonal hypersurface of dimension $n = 2d$ and prime degree $m > 3$ with twist $\mathbf{c}$ over $k = \mathbb{F}_q$. Suppose $\mathcal{V}_k$ is of Hodge–Witt type. Then the exponent of q in the residue $C_{\mathcal{V}}(d)$ in the Lichtenbaum–Milne formula (Conjecture 0.2) is correct, that is,

$$\chi(\mathcal{V}, \mathcal{O}, d) = \sum_{i=0}^{d} (d-i) h^{i,n-i}(\mathcal{V}) = w_{\mathcal{V}}(d).$$

Furthermore, assume that the complexes $\mathbb{Z}(r)$ exist and that the cycle map is surjective. Then

$$\chi(\mathcal{V}_k, \mathbb{Z}(d)) = \frac{(-1)^d q^{-d(d+1)/2} \prod_{i=1}^{d} (q^i - 1)^{-2}}{B^d(\mathcal{V}_k) \cdot m^{\delta_d(\mathcal{V}_k)}} \in \mathbb{Q}$$

where $\delta_d(\mathcal{V}_k)$ is as defined in Chapter 7.

Proof: This follows from Theorem 7.3 and Theorem 8.7. □

Now we consider an odd-dimensional diagonal hypersurface $\mathcal{V} = \mathcal{V}_m^n(\mathbf{c})$ over $k = \mathbb{F}_q$. Let $n = 2d + 1$ and $m > 3$ be prime. Then for each integer r, $0 \le r \le d$, the Tate conjecture is obviously true as $H^{2r}(\mathcal{V}_{\bar{k}}, \mathbb{Q}_\ell(r))$ has dimension 1.

Now we are interested in the special value of $Z(\mathcal{V}_k, q^{-s})$ at $s = r$. We obtain from Theorem 7.6,

$$\lim_{s \to r} Z(\mathcal{V}, q^{-s})(1 - q^{r-d}) = Q(\mathcal{V}, q^{-r}) \prod_{\substack{i=0 \\ i \neq r}}^{2d} (1 - q^{i-r})^{-1}$$

$$= q^{-w_{\mathcal{V}}(r)} D^d(\mathcal{V}_k) \prod_{\substack{i=0 \\ i \neq r}}^{2d} (1 - q^{i-d})^{-1}.$$

Now we compare this with the Lichtenbaum–Milne conjecture.

THEOREM 8.11

Let $\mathcal{V} = \mathcal{V}_n^m(\mathbf{c})$ be a diagonal hypersurface of dimension $n = 2d + 1$ and prime degree $m > 3$ with twist $\mathbf{c}$ over $k = \mathbb{F}_q$. Assume that for any r, $0 \le r \le d$, the complexes $\mathbb{Z}(r)$ exist and the cycle map is surjective. Then $\mathcal{V}_k$ satisfies the Lichtenbaum–Milne formula, that is, the exponent of q is

$$\chi(\mathcal{V}, \mathcal{O}, r) = \sum_{i=0}^{r} (d - i) h^{i,n-i}(\mathcal{V}) = w_{\mathcal{V}}(r),$$

and $\chi(\mathcal{V}_k, \mathbb{Z}(r))$ is given explicitly by

$$\chi(\mathcal{V}_k, \mathbb{Z}(r)) = \frac{D^r(\mathcal{V}_k)}{(-1)^r q^{r(r+1)/2} \prod_{i=1}^{r}(q^i - 1)^2 \cdot \prod_{j=1}^{2d-2r}(1 - q^{r+j})} \in \mathbb{Q}.$$

Proof: This follows from Theorem 0.1 of Milne and Theorem 7.6. □

9 Remarks, observations and open problems

We conclude with various remarks, and in particular with some observations on the cases about which we have fewer results.

9.1 The case of composite m

Many of the results obtained in this paper are restricted to diagonal hypersurfaces of *prime* degree m. This restriction is not a subtle one, but rather a technical one. In fact, we have some rudimentary results for diagonal hypersurfaces of composite degree m.

Let $\mathcal{V}_k = \mathcal{V}_k(\mathbf{c})$ be a diagonal hypersurface of degree m and dimension n with twist $\mathbf{c}$ defined over $k = \mathbb{F}_q$.

The Picard numbers. About the combinatorial Picard numbers for $\mathcal{V}_k$ of even dimension $n = 2d$, we note that the second assertion of Proposition 5.5 no longer holds; indeed, in some cases there are twists $\mathbf{c}$ satisfying

$$\rho_d(\mathcal{V}_k) > \rho_d(\mathcal{X}_k)$$

where

$$\rho_d(\mathcal{V}_k) = \#\{\mathbf{a} \in \mathfrak{A}_n^m \mid \chi(\mathbf{c}^{\mathbf{a}}) = j(\mathbf{a})/q^d\}.$$

For instance, take $(m, n) = (4, 2)$, $p = 5$. Then $\rho_1(\mathcal{X}_k) = 8$. Now choose twists $\mathbf{c} = (4, 4, 1, 1)$ (resp. $\mathbf{c} = (3, 4, 2, 1)$). Then $\rho_1(\mathcal{V}_k) = 16$ (resp. (10)). As another example, take $(m, n) = (10, 4)$, $p = 11$. Then $\rho_2(\mathcal{X}_k) = 4061$, but for a twist $\mathbf{c} = (10, 10, 10, 1, 1, 1)$, we have $\rho_2(\mathcal{V}_k) = 5218$.

We computed the combinatorial Picard numbers for various twists in a number of cases (the results are tabulated in Section A.3). In each case, we pick a non-primitive root g modulo p, and consider twists $\mathbf{c} = (c_0, c_1, \cdots, c_{n+1})$ where each component c_i is of the form g^j, $1 \leq j \leq m - 1$. Based on our computations, we observe the following facts:

1. **Extreme twists:** In the composite case, our definition of extreme twists does not work. Instead, we simply say a twist is extreme if the combinatorial Picard number of the resulting variety is 1. Such twists seem to be very hard to find. In our computations, they occurred only for $n = 2$ and $m = 9, 14$, or 15, and all of them were equivalent to twists

of the form $(c_0, 1, 1, \ldots, 1)$ with $c_0 \neq 1$. For most values of m and n, we found no extreme twists at all.

2. **The inequality $\rho_d(\mathcal{V}_k) \leq \rho(\mathcal{X}_k)$**: As we pointed out above, this does not hold for general m. On the other hand, our computations suggest that it does hold when m is odd.

3. **Stable Picard numbers**: in the case of prime degree, the stable combinatorial Picard number is equal to the combinatorial Picard number of the Fermat hypersurface. In the case of composite degree, this seems never to happen: the combinatorial Picard number seems to be always significantly smaller than the stable combinatorial Picard number.

For more examples, see Section A.3 in the appendix.

The norms. We have computed the norms of algebraic numbers $1 - \mathcal{J}(\mathbf{c}, \mathbf{a})/q^d$ for some selected twists $\mathbf{c}$. Here we record some partial results, and observations based on our computations.

1. If m is odd and $m = m_0^r$ where m_0 is an odd prime and $r \geq 2$, then, as pointed out above, the Iwasawa–Ihara conjecture holds, and we have

$$\mathcal{J}(\mathbf{c}, \mathbf{a}) \equiv 1 \pmod{(1 - \zeta)}$$

where ζ is an m-th root of unity. Consequently, for any r, $0 \leq r \leq n$,

$$\mathrm{Norm}_{L/\mathbb{Q}}(1 - \frac{\mathcal{J}(\mathbf{c}, \mathbf{a})}{q^d}) \equiv 0 \pmod{m_0}$$

and higher powers of m_0 will occur when the twist is trivial.

2. For general m, there seems to be no general pattern for the powers of the prime divisors of m in the norm. See the tables for various examples of this.

3. For arbitrary m and $\mathbf{c}$, the p-part of the norms satisfies

$$\mathrm{Norm}_{L/\mathbb{Q}}(1 - \frac{\mathcal{J}(\mathbf{c}, \mathbf{a})}{q^r}) \in \frac{1}{q^{w(r)}}\mathbb{Z} \quad \text{for any } r, \ 0 \leq r \leq n$$

where $w(r)$ is as in Theorem 6.2.

4. For arbitrary m, $\mathbf{c}$ and $n = 2d$, we have

$$q^{w(r)} \mathrm{Norm}_{L/\mathbb{Q}}(1 - \frac{\mathcal{J}(\mathbf{c}, \mathbf{a})}{q^d})$$

is a square up to a factor involving only the primes dividing m.

Examples: We give a number of examples of norms for selected twisted Jacobi sums of composite degree.

Each table below records the numbers

$$q^{w(d)} \, \mathrm{Norm}_{L/\mathbb{Q}}\left(1 - \frac{\mathfrak{J}(\mathbf{c}, \mathbf{a})}{q^d}\right)$$

for various values of m, n, p, and $\mathbf{a}$, and two different twists. Supersingular characters are indicated by $*$.

(a) Let $m = 6$, $n = 2$, $p = 7$, and let

$$\mathbf{c}_1 = (2, 1, 1, 1) \qquad \text{and} \qquad \mathbf{c}_2 = (2, 3, 5, 1).$$

$\mathbf{a}$	$w(r)$	$\mathbf{c}_1$	$\mathbf{c}_2$
$*(1, 1, 5, 5)$	0	3	3
$*(1, 2, 4, 5)$	0	1	2^2
$(1, 1, 1, 3)$	1	$2^2 \cdot 3$	$2^2 \cdot 3$
$(3, 1, 1, 1)$	1	3^3	3^3
$(1, 1, 2, 2)$	1	5^2	3^3
$(2, 2, 1, 1)$	1	1	3

(b) Let $m = 9$, $n = 2$, $p = 19$, and let

$$\mathbf{c}_1 = (2, 1, 1, 1) \qquad \text{and} \qquad \mathbf{c}_2 = (2, 3, 5, 1).$$

$\mathbf{a}$	$w(r)$	$\mathbf{c}_1$	$\mathbf{c}_2$
$*(1, 1, 8, 8)$	0	3	3
$*(1, 4, 6, 7)$	0	3	3
$(1, 1, 1, 6)$	2	3	3^3
$(6, 1, 1, 1)$	2	3^7	3
$(1, 1, 2, 5)$	2	3	3
$(5, 1, 1, 2)$	2	$3 \cdot 17^2$	$3 \cdot 37^2$
$(1, 2, 3, 3)$	1	3	3^3

(c) Let $m = 4$, $n = 4$, $p = 5$, and let

$$\mathbf{c}_1 = (2,1,1,1,1,1) \qquad \text{and} \qquad \mathbf{c}_2 = (2,3,5,1,1,1).$$

$\mathbf{a}$	$w(r)$	$\mathbf{c}_1$	$\mathbf{c}_2$
$*(1,1,1,3,3,3)$	0	2	2
$*(1,1,2,2,3,3)$	0	2	2
$*(2,1,1,2,3,3)$	0	2^2	2
$(1,3,3,3,3,3)$	1	2	$2 \cdot 3^2$
$(3,1,3,3,3,3)$	1	$2 \cdot 3^2$	$2 \cdot 3^2$
$(1,1,1,1,2,2)$	1	2	$2 \cdot 3^2$
$(2,1,1,1,2,1)$	1	2^2	$2 \cdot 3^2$

(d) Let $m = 6$, $n = 6$, $p = 7$, and let

$$\mathbf{c}_1 = (2,1,1,1,1,1,1,1) \qquad \text{and} \qquad \mathbf{c}_2 = (2,3,5,1,1,1,1,1).$$

$\mathbf{a}$	$w(r)$	$\mathbf{c}_1$	$\mathbf{c}_2$
$*(1,1,1,2,4,5,5,5)$	0	1	1
$*(1,1,2,1,4,5,5,5)$	0	1	3
$*(1,1,1,3,3,5,5,5)$	0	3	3
$(1,5,5,5,5,5,5,5)$	2	$2^6 \cdot 3$	$2^6 \cdot 3$
$(5,1,5,5,5,5,5,5)$	2	$3 \cdot 5^2$	3^3
$(1,1,3,5,5,5,5,5)$	1	3	3^3
$(2,1,2,5,5,5,5,5)$	1	1	3
$(5,1,1,4,4,5,5,5)$	1	5^2	5^2
$(1,1,1,1,1,1,2,4)$	2	11^2	11^2
$(2,1,1,1,1,1,2,3)$	2	13^2	2^2
$(1,1,1,3,3,3,3,3)$	1	$2^2 \cdot 3$	$2^2 \cdot 3$
$(1,2,2,2,2,2,3,4)$	1	1	1
$(1,2,3,2,2,2,2,4)$	1	1	$2^2 \cdot 3$
$(2,2,2,2,2,2,3,3)$	1	5^2	5^2

(e) Let $m = 15$, $n = 6$, $p = 31$, and let

$$\mathbf{c}_1 = (2,1,1,1,1,1,1,1) \quad \text{and} \quad \mathbf{c}_2 = (2,3,5,1,1,1,1,1).$$

a	$w(r)$	$\mathbf{c}_1$	$\mathbf{c}_2$
$*(1,1,1,1,14,14,14,14)$	0	5^2	0
$(1,5,14,14,14,14,14)$	5	$29^2 \cdot 509^2$	$29^2 \cdot 509^2$
$(1,14,5,14,14,14,14,14)$	5	$29^2 \cdot 509^2$	239^2
$(1,8,11,14,14,14,14,14)$	4	$2^8 \cdot 421^2$	1381^2
$(8,1,11,14,14,14,14,14)$	4	1381^2	$3^4 \cdot 479^2$
$(1,3,3,7,8,10,14,14)$	1	5^4	5^2
$(1,3,3,7,7,11,14,14)$	4	5^4	$2^8 \cdot 5^2 \cdot 29^2$
$(3,1,3,7,7,11,14,14)$	4	$2^8 \cdot 5^2 \cdot 29^2$	2281^2
$(1,3,3,6,8,11,14,14)$	1	29^2	2^8
$(14,3,3,6,8,11,14,1)$	1	3^4	29^2

9.2 The plus norms

Let $\mathcal{V}_A$ be a twisted Fermat motive of dimension n and degree m over $k = \mathbb{F}_q$. We should compute also the plus norms

$$\mathrm{Norm}_{L/\mathbb{Q}}(1 + \mathcal{J}(\mathbf{c},\mathbf{a})/q^r), \quad 0 \leq r \leq n$$

as done in Yui [Yu94].

PROPOSITION 9.1
Assume that $\mathcal{V}_A$ is of Hodge–Witt type. Then the following assertions hold:

1. *Let m and n be arbitrary. Then for any r, $0 \leq r \leq n$, the p-part of the norm is equal to $q^{w\nu_A(r)}$.*

2. *Assume that m is prime, and $\mathbf{c}$ is extreme.*

   (a) *Assume that n is even. Then for any r, $0 \leq r \leq n$,*

   $$q^{w\nu_A(r)} \cdot \mathrm{Norm}_{L/\mathbb{Q}}(1 + \mathcal{J}(\mathbf{c},\mathbf{a})/q^r) = B_+^r(\mathcal{V}_A)$$

   where $B_+^r(\mathcal{V}_A)$ is a positive integer relatively prime to m (but not necessarily prime to p).

(b) Assume that n is odd. Then for any r, $0 \le r \le n$,

$$q^{w\nu_A(r)} \cdot \operatorname{Norm}_{L/\mathbb{Q}}(1 + \mathfrak{J}(\mathbf{c}, \mathbf{a})/q^r) = D^r_+ \cdot m$$

where D^r_+ is a positive integer relatively prime to m (but not necessarily prime to p).

Proof: (1) This is true by the same reasoning as for the p-part in Theorem 6.2.

(2) Let $\mathfrak{J}_2(\mathbf{c}, \mathbf{a})$ denote a twisted Jacobi sum relative to $k_2 = \mathbb{F}_{q^2}$. Note that $\mathfrak{J}_2(\mathbf{c}, \mathbf{a}) = \mathfrak{J}(\mathbf{c}, \mathbf{a})^2$. Then

$$\operatorname{Norm}_{L/\mathbb{Q}}(1 - \mathfrak{J}_2(\mathbf{c}, \mathbf{a})/q^{2r})$$
$$= \operatorname{Norm}_{L/\mathbb{Q}}(1 - \mathfrak{J}(\mathbf{c}, \mathbf{a})/q^r) \cdot \operatorname{Norm}_{L/\mathbb{Q}}(1 + \mathfrak{J}(\mathbf{c}, \mathbf{a})/q^r).$$

Observe that both Jacobi sums $\mathfrak{J}_2(\mathbf{c}, \mathbf{a})$ and $\mathfrak{J}(\mathbf{c}, \mathbf{a})$ satisfy the congruence of Proposition 1.7:

$$\mathfrak{J}_2(\mathbf{c}, \mathbf{a}) \equiv \mathfrak{J}(\mathbf{c}, \mathbf{a}) \equiv 1 \pmod{(1 - \zeta)}.$$

Moreover, the assertions of Theorem 6.2 are also valid for the minus norm for $\mathfrak{J}_2(\mathbf{c}, \mathbf{a})$, that is,

$$q^{2\,w\nu_A(r)} \cdot \operatorname{Norm}_{L/\mathbb{Q}}(1 - \mathfrak{J}_2(\mathbf{c}, \mathbf{a})/q^{2r}) = \begin{cases} B^r_2(\mathcal{V}_A) \cdot m & \text{if } n \text{ is even} \\ D^r_2(\mathcal{V}_A) & \text{if } n \text{ is odd} \end{cases}$$

where $B^r_2(\mathcal{V}_A)$ and $D^r_2(\mathcal{V}_A)$ denote the positive integers defined as in Theorem 6.2 for the twisted Fermat motive $\mathcal{V}_A$ corresponding to $\mathfrak{J}_2(\mathbf{c}, \mathbf{a})$.

If n is even (resp. odd), the minus norms for $\mathfrak{J}_2(\mathbf{c}, \mathbf{a})$ and $\mathfrak{J}(\mathbf{c}, \mathbf{a})$ have the m-adic order 1 (resp. 0). Therefore, the norm identity above yields the assertion on the plus norm. $\square$

Proposition 9.1 gives a way of deducing the norms for Jacobi sums $\mathfrak{J}_2(\mathbf{c}, \mathbf{a})$ relative to the quadratic extension $k_2 = \mathbb{F}_{q^2}$ from those relative to k.

9.3   Further questions

Our investigation raises a number of questions that seem worthy of further study.

1. Conjecture 1.9 ought to be studied further.

2. We have made no use of the varieties constructed in Proposition 3.2. It might be worthwhile to investigate their geometry.

3. The question of the existence and frequency of extreme twists also seems interesting.

4. When $\mathcal{V}$ is supersingular, one should be able to compute the invariants $\alpha_d(\mathcal{V})$. This is equivalent to the determination of the structure of the unipotent formal groups attached to $\mathcal{V}$. In particular, one would like to know the number of copies of $\hat{\mathbb{G}}_a$'s occurring in them (cf. Proposition 8.4).

5. At several points, we make the assumption that the motives and/or the hypersurfaces with which we work are of Hodge–Witt type. At the global (non-motivic) level, this is quite a heavy assumption: except for a small list of exceptions, it restricts us to the ordinary case. (See Remark 3.15.) Suwa [Su] has just announced that he can extend several of the results in Chapter 8 by eliminating this assumption.

6. The obvious next step is to go on to study arithmetical properties of diagonal hypersurfaces over number fields. Some investigation along these lines has already been started. Pinch and Swinnerton-Dyer [PS91] considered diagonal quartic surfaces over $\mathbb{Q}$ addressing these questions, in which case the L-series was expressed in terms of Hecke L-series with Grossencharacters over $\mathbb{Q}(i)$. Harrison [Ha92] analysed the special value of the L-function of a diagonal quartic surface over $\mathbb{Q}$ proving the Block–Kato conjecture (see [BK90]) on Tamagawa numbers for Grossencharacters over $\mathbb{Q}(i)$.

More generally, Shuji Saito [Sa89] considered arithmetic surfaces and proved among other things the finiteness of the Brauer group. Also Parshin [Pa83] discussed special values of zeta–functions of function fields over number fields.

As an attempt to generalise the results for diagonal quartic surfaces by Pinch and Swinnerton-Dyer and Harris, Goto, Gouvêa and Yui [GGY] have started considering the above questions for weighted diagonal K3–surfaces defined over number fields.

A   Tables

In this appendix, we collect many of the results of our computations. After a short discussion of what was involved in the computations, we begin by giving examples of what we have called the "twisted Fermat motives" and work out their invariants. Section A.3 gives combinatorial Picard numbers (both stable and over the prime field $k = \mathbb{F}_p$). Section A.4 deals with the "Brauer numbers" of twisted Fermat motives, which are computed in a number of cases. Finally, the last section presents a few calculations of the "Brauer number" of a diagonal hypersurface, achieved by computing the numbers in question for a representative set of motives and then working out their multiplicity.

A.1   A note on the computations

The computations were done mostly at Colby College, using a variety of computer equipment and several software packages. The preliminary computations were done with *Mathematica* on a Macintosh Quadra 950. These provided us with a basic outline of what the results should look like. The final computations were done with C programs using the PARI library for number-theoretic functions and infinite-precision arithmetic. The portability of both C programs and the PARI library allowed us to run the programs on a VAX running Ultrix, a SparcStation, an Intel-based machine running OS/2, and on a Macintosh Quadra 950. The software used included the PARI library (by H. Cohen et. al.), the Gnu C Compiler (including the OS/2 port by Eberhard Mattes), and the MPW C compiler on the Macintosh.

Most of the techniques used to do the computations were straightforward, requiring mostly time and large amounts of computer memory. The full computations required several months to be completed. Gaps in the tables above reflect cases where we were unable to complete the computations.

The more ambitious computations were greatly helped by a grant from Colby College, which allowed us to upgrade one of our machines. In writing the programs themselves, we received a great deal of help from a student at Colby College, Lynette Millett, who was our research assistant over one summer and did much of the initial programming. We also thank Henri Cohen for his help with the PARI library.

The fundamental strategy for organizing the computations was exploiting the motivic decomposition, and in particular the fact that two Fermat motives are isomorphic when one character in the first is a permutation of a character in the second. Hence, for each m and n, we began by generating a representative for each of the Fermat motives. We then eliminated from the list any "duplicates" (i.e., any character whose motive was isomorphic to one already represented). This gave us our basic list of characters.

Many of the quantities we were looking for were independent of the twist, and hence could be computed directly from our minimal list of characters. When a twist was present and relevant, we took it into account by breaking the isomorphism class of motives (given by permutation) into subclasses for which the twisted motives were isomorphic. For example, for a twist of the form $(c_0, 1, 1, \ldots, 1)$ with $c_0 \neq 1$, we only need to consider how many permutations have a certain entry in position 0, since other entries can be permuted without changing the isomorphism class.

After computing our data for each isomorphism class, one needs to worry about the multiplicity of each class. This is simply a matter of counting permutations, except for cases when a motive is "self-isomorphic", i.e., when there is a permutation of $\mathbf{a}$ which is *also* a multiple of $\mathbf{a}$. These occur fairly often, in fact.

Most of the computations are then straightforward. Jacobi sums were computed by using their expression in terms of Gauss sums (see Proposition 1.2), which were computed directly. (This is by far the most time-consuming portion of the computations.) Norms were also computed in a straightforward fashion. In every case, we chose to work over the prime field, and therefore under the assumption that $p \equiv 1 \pmod{m}$. This simplifies the computation quite significantly, and has two further advantages: first, in many cases the Tate conjecture is known to hold (see Proposition 5.17); second, determining which characters are supersingular turns out not to depend on the specific prime p (i.e., the answer is the same for all $p \equiv 1 \pmod{m}$).

For the most part, the tables below represent only a sampling of our output, chosen to exemplify the various sorts of phenomena we observed.

A.2   Twisted Fermat motives and their invariants

We will describe some of the twisted Fermat motives of arbitrary degree and even dimension for selected twists $\mathbf{c}$. If $\mathbf{c} = \mathbf{1}$, then of course $\chi(\mathbf{c^{ta}}) = 1$ for all $t \in (\mathbb{Z}/m\mathbb{Z})^\times$; these will not be listed in the tables. The tables list the information that "makes up" twisted Fermat motives with twists $\mathbf{c}$, and then also list their invariants, e.g., the (non-zero) Hodge numbers and the Betti number. We use ζ for $e^{2\pi i/m}$ in each case. Roots of unity are normalized in terms of a basis of $\mathbb{Z}[\zeta]$.

(a) Let $(m, n) = (8, 4)$, and take a character $\mathbf{a} = (1, 1, 2, 3, 4, 5) \in \mathfrak{A}_4^8$. Choose $q = p = 17$. Consider two twists

$$\mathbf{c}_1 = (11, 15, 13, 10, 9, 1) \quad \text{and} \quad \mathbf{c}_2 = (5, 3, 3, 1, 1, 1).$$

To describe the twisted Fermat motive $\mathcal{V}_A$ we list the character and its scalings, the length of $t\mathbf{a}$, and the corresponding roots of unity $\bar{\chi}(\mathbf{c}^{t\mathbf{a}})$. For each motive we list the (non-zero) Hodge numbers and the Betti number.

t	$t\mathbf{a}$	$\bar{\chi}(\mathbf{c}_1^{t\mathbf{a}})$	$\bar{\chi}(\mathbf{c}_2^{t\mathbf{a}})$	$\|t\mathbf{a}\|$	$h^{i,j}$
1	$(1, 1, 2, 3, 4, 5)$	ζ^2	1	1	$h^{1,3} = h^{3,1} = 1$
3	$(3, 3, 6, 1, 4, 7)$	$-\zeta^2$	1	2	$h^{2,2} = 2$
5	$(5, 5, 2, 7, 4, 1)$	ζ^2	1	2	$B_4 = 4$
7	$(7, 7, 6, 5, 4, 3)$	$-\zeta^2$	1	3	

(b) Let $(m, n) = (15, 4)$. Take a character $\mathbf{a} = (1, 1, 1, 2, 2, 8) \in \mathfrak{A}_4^{15}$. Let $q = p = 31$. Choose a twist $\mathbf{c}_1 = (17, 19, 27, 3, 9, 1)$, and its permutation $\mathbf{c}_2 = (1, 9, 3, 27, 19, 17)$. Then the twisted Fermat motives $\mathcal{V}_A$ and their invariants are as follows:

t	$t\mathbf{a}$	$\bar{\chi}(\mathbf{c}_1^{t\mathbf{a}})$	$\bar{\chi}(\mathbf{c}_2^{t\mathbf{a}})$	$\|t\mathbf{a}\|$	$h^{i,j}$
1	$(1, 1, 1, 2, 2, 8)$	$-\zeta^5 - 1$	ζ^2	0	$h^{0,4} = h^{4,0} = 2$
2	$(2, 2, 2, 4, 4, 1)$	ζ^5	ζ^4	0	$h^{1,3} = h^{3,1} = 2$
4	$(4, 4, 4, 8, 8, 2)$	$-\zeta^5 - 1$	$\zeta^7 - \zeta^5 + \zeta^4 - \zeta^3 + \zeta - 1$	1	$B_4 = 8$
7	$(7, 7, 7, 14, 14, 11)$	$-\zeta^5 - 1$	$-\zeta^7 + \zeta^6 - \zeta^4 + \zeta^3 - \zeta^2 + 1$	3	
8	$(8, 8, 8, 1, 1, 4)$	ζ^5	ζ	1	
11	$(11, 11, 11, 7, 7, 13)$	ζ^5	ζ^7	3	
13	$(13, 13, 13, 11, 11, 14)$	$-\zeta^5 - 1$	$-\zeta^6 - \zeta$	4	
14	$(14, 14, 14, 13, 13, 7)$	ζ^5	$-\zeta^7 + \zeta^5 - \zeta^4 - \zeta + 1$	4	

Observe that motives are not invariant under permutation of twists, though the geometric invariants remain the same. The slight difference in the roots of unity that one sees in this table turns out to have a serious impact on the arithmetic (e.g., on the combinatorial Picard number, and of course on the special value of the zeta-function). Both motives are ordinary (since we have chosen $p \equiv 1 \pmod{15}$; they are not supersingular.

(c) Let $(m, n) = (25, 4)$. Take a character $\mathbf{a} = (1, 1, 2, 3, 4, 14) \in \mathfrak{A}_4^{25}$. Let $q = p = 101$. Choose twists $\mathbf{c}_1 = (8, 4, 2, 4, 8, 1)$ and $\mathbf{c}_2 = (7, 7, 1, 1, 1, 1)$. Notice that $\mathbf{c}_1^{\mathbf{a}} = 2^{25}$ is a 25-th power.

t	$t\mathbf{a}$	$\bar{\chi}(\mathbf{c}_1^{t\mathbf{a}})$	$\bar{\chi}(\mathbf{c}_2^{t\mathbf{a}})$	$\|t\mathbf{a}\|$	$h^{i,j}$
1	$(1, 1, 2, 3, 4, 14)$	1	ζ^7	0	$h^{0,4} = h^{4,0} = 2$
2	$(2, 2, 4, 6, 8, 3)$	1	ζ^{14}	0	$h^{1,3} = h^{3,1} = 4$
3	$(3, 3, 6, 9, 12, 17)$	1	$-\zeta^{16} - \zeta^{11} - \zeta^6 - \zeta$	1	$h^{2,2} = 8$
4	$(4, 4, 8, 12, 16, 6)$	1	ζ^3	1	$B_4 = 20$
6	$(6, 6, 12, 18, 24, 9)$	1	ζ^{17}	2	
7	$(7, 7, 14, 21, 3, 23)$	1	$-\zeta^{19} - \zeta^{14} - \zeta^9 - \zeta^4$	2	
8	$(8, 8, 16, 24, 7, 12)$	1	ζ^6	2	
9	$(9, 9, 18, 2, 11, 1)$	1	ζ^{13}	1	
11	$(11, 11, 22, 8, 19, 4)$	1	ζ^2	2	
12	$(12, 12, 24, 11, 23, 18)$	1	ζ^9	3	
13	$(13, 13, 1, 14, 2, 7)$	1	ζ^{16}	1	
14	$(14, 14, 3, 17, 6, 21)$	1	$-\zeta^{18} - \zeta^{13} - \zeta^8 - \zeta^3$	2	
16	$(16, 16, 7, 23, 14, 24)$	1	ζ^{12}	3	
17	$(17, 17, 9, 1, 18, 13)$	1	ζ^{19}	2	
18	$(18, 18, 11, 4, 22, 2)$	1	ζ	2	
19	$(19, 19, 13, 7, 1, 16)$	1	ζ^8	2	
21	$(21, 21, 17, 13, 9, 19)$	1	$-\zeta^{17} - \zeta^{12} - \zeta^7 - \zeta^2$	3	
22	$(22, 22, 19, 16, 13, 8)$	1	ζ^4	3	
23	$(23, 23, 21, 19, 17, 22)$	1	ζ^{11}	4	
24	$(24, 24, 23, 22, 21, 11)$	1	ζ^{18}	4	

Notice that, as expected, the motive attached to the first twist is in fact isomorphic to the (untwisted) Fermat motive.

The Hodge and Newton polygons have slopes $\{0, 1, 2, 3, 4\}$ with respective multiplicities $2, 4, 8, 4, 2$. These motives are ordinary, but not supersingular.

If we take $q = p \equiv 11, 16 \pmod{25}$, then $p^5 \equiv 1 \pmod{25}$, and both motives become supersingular, with the Newton polygon having the pure slope 2.

There are many twists which give rise to motives that are isomorphic to the one given by $\mathbf{c}_1$, e.g., $\mathbf{c} = (2, 4, 4, 4, 8, 1)$, $\mathbf{c} = (8, 2, 8, 2, 8, 1)$, $\mathbf{c} = (4, 8, 2, 4, 8, 1)$ and others. This is a general phenomenon: given a character $\mathbf{a}$, many twists produce isomorphic motives $\mathcal{V}_A$. Conversely, given the twist $\mathbf{c}$, many of its characters yield isomorphic motives. This is one of the crucial simplifying observations for the computations. When one wants a computation for the diagonal hypersurface $\mathcal{V}$, one needs to identify which characters produce isomorphic motives for the given twist, and then work out the corresponding multiplicities.

(d) Let $(m, n) = (7, 6)$. Choose a character $\mathbf{a} = (1, 1, 1, 2, 2, 2, 2, 3) \in \mathfrak{A}_6^7$. Let $q = p = 29$. We work with the twists

$$\mathbf{c}_1 = (1, 2, 3, 3, 4, 4, 5, 6) \quad \text{and} \quad \mathbf{c}_2 = (5, 6, 1, 2, 3, 4, 3, 4).$$

Once again, these are permutations of each other. The geometric invariants of the two motives are the same (since they are twists of the same Fermat motive), but notice that the roots of unity determined by the twist are completely different. The twisted Fermat motives are described as follows:

t	$t\mathbf{a}$	$\bar{\chi}(\mathbf{c}_1^{t\mathbf{a}})$	$\bar{\chi}(\mathbf{c}_2^{t\mathbf{a}})$	$\|\mathbf{a}\|$	$h^{i,j}$
1	$(1, 1, 1, 2, 2, 2, 2, 3)$	ζ^5	ζ^3	1	$h^{1,5} = h^{5,1} = 1$
2	$(2, 2, 2, 4, 4, 4, 4, 6)$	ζ^3	ζ^6	3	$h^{2,4} = h^{4,2} = 1$
3	$(3, 3, 3, 6, 6, 6, 6, 2)$	ζ	ζ^2	4	$h^{3,3} = 2$
4	$(4, 4, 4, 1, 1, 1, 1, 5)$	ζ^6	ζ^5	2	$B_6 = 6$
5	$(5, 5, 5, 3, 3, 3, 3, 1)$	ζ^4	ζ	3	
6	$(6, 6, 6, 5, 5, 5, 5, 4)$	ζ^2	ζ^4	5	

These two motives are ordinary, since $p \equiv 1 \pmod 7$. They are not supersingular.

If we choose $q = p \equiv 2$ or $4 \pmod 7$, then $\mathcal{V}_A$ is of Hodge–Witt type but not ordinary, as the Newton polygon has slopes $\{2, 4\}$ with each of multiplicity 3, while the Hodge polygon has slopes $\{1, 2, 3, 4, 5, 6\}$ with each of multiplicity

1. If we choose $q = p \equiv 3$ or 5 (mod 7), then $\mathcal{V}_A$ is supersingular, but not ordinary.

(e) Let $(m, n) = (11, 6)$. Choose a character $\mathbf{a} = (1, 1, 1, 1, 2, 3, 6, 7) \in \mathfrak{A}_6^{11}$. Let $q = p = 23$ (so that, once again, $p \equiv 1$ (mod 11)). Choose twists

$$\mathbf{c}_1 = (2, 1, 2, 1, 2, 1, 2, 1) \qquad \text{and} \qquad \mathbf{c}_2 = (1, 1, 1, 1, 2, 2, 2, 2).$$

Once again, these are permutations of each other. Then the twisted Fermat motives $\mathcal{V}_A$ are described as follows:

t	ta	$\bar{\chi}(\mathbf{c}_1^{ta})$	$\bar{\chi}(\mathbf{c}_2^{ta})$	$\|\mathbf{a}\|$	$h^{i,j}$
1	$(1, 1, 1, 1, 2, 3, 6, 7)$	ζ^2	ζ^8	1	$h^{1,5} = h^{5,1} = 2$
2	$(2, 2, 2, 2, 4, 6, 1, 3)$	ζ^4	ζ^5	1	$h^{2,4} = h^{4,2} = 1$
3	$(3, 3, 3, 3, 6, 9, 7, 10)$	ζ^6	ζ^2	3	$h^{3,3} = 3$
4	$(4, 4, 4, 4, 8, 1, 2, 6)$	ζ^8	ζ^{10}	2	$h^{i,j} = 0$ otherwise
5	$(5, 5, 5, 5, 10, 4, 8, 2)$	ζ^{10}	ζ^7	3	$B_6 = 10$
6	$(6, 6, 6, 6, 1, 7, 3, 9)$	ζ	ζ^4	3	
7	$(7, 7, 7, 7, 3, 10, 9, 5)$	ζ^3	ζ	4	
8	$(8, 8, 8, 8, 5, 2, 4, 1)$	ζ^5	ζ^9	3	
9	$(9, 9, 9, 9, 7, 5, 10, 8)$	ζ^7	ζ^6	5	
10	$(10, 10, 10, 10, 9, 8, 5, 4)$	ζ^9	ζ^3	5	

Both motives are, of course, ordinary. In both cases, if we change our prime so that $q = p \equiv \{2, 6, 7, 8\}$ (mod 11), then the $\mathcal{V}_A$ become supersingular; while if we choose $q = p \equiv \{3, 4, 5, 9\}$ (mod 11), then $\mathcal{V}_A$ are neither ordinary, nor supersingular, nor of Hodge–Witt type.

A.3   Picard numbers of $\mathcal{V} = \mathcal{V}_n^m(\mathbf{c})$

The tables in this section give the combinatorial Picard numbers and stable combinatorial Picard numbers of diagonal hypersurfaces. In the cases covered by Proposition 5.17, the Tate conjecture holds, so that the numbers we compute are the geometric (rather than merely combinatorial) Picard numbers. We will omit the word "combinatorial" to simplify the exposition.

For each dimension n, we begin by computing the stable Picard numbers and the Picard numbers for the Fermat hypersurfaces of dimension n and degree m, where m ranges from 4 to the largest value we could handle in our computations. We then compute Picard numbers of twisted hypersurfaces of small degree, for many different twists (the tables give only a selection of our results).

A.3.1 Dimension 2. (a) We compute the Picard numbers and the stable Picard numbers of diagonal hypersurfaces with trivial twist of dimension $n = 2$ and degree m where $4 \leq m \leq 49$ choosing $k = \mathbb{F}_q$ with $q = p \equiv 1 \pmod{m}$. The Tate conjecture holds in every case.

m	p	$\rho_1(\mathcal{X}_k)$	$\bar{\rho}_1(\mathcal{V})$	m	p	$\rho_1(\mathcal{X}_k)$	$\bar{\rho}_1(\mathcal{V})$
4	5	8	20	27	109	1951	2143
5	11	37	37	28	29	1064	2972
6	13	26	86	29	59	2269	2269
7	29	91	91	30	31	1226	5630
8	17	128	176	31	311	2611	2611
9	19	169	217	32	97	1352	3416
10	11	98	362	33	67	2977	3217
11	23	271	271	34	103	1538	3938
12	13	152	644	35	71	3367	3367
13	43	397	397	36	73	1856	5516
14	29	518	806	37	149	3781	3781
15	31	547	835	38	191		4862
16	17	296	872	39	79	4219	4507
17	103	721	721	40	41	2168	6224
18	37	386	1658	41	83	4681	4681
19	191	919	919	42	43	2450	10418
20	41	1028	1988	43	173	5167	5167
21	43	1141	1573	44	89		6380
22	23	602	1742	45	271		6205
23	47	1387	1387	46	47		6998
24	73	1136	3080	47	283	6211	6211
25	51	1657	1657	48	97		9200
26	53	1802	2378	49	197		6769

Our computational results are consistent with a closed formula for the stable Picard number $\bar{\rho}_1(\mathcal{V})$ due to Shioda [Sh82a] and Aoki and Shioda [AS83]. It is given by

$$\bar{\rho}_1(\mathcal{V}) = 1 + 3(m-1)(m-2) + \delta_m + 48(m/2)^* + 24(m/3)^* + 24\varepsilon(m),$$

where

- δ_m is 0 or 1 depending on m being odd or even,

- if $\alpha \in \mathbb{Q}$ is a rational number, then $(\alpha)^* = \alpha$ if $\alpha \in \mathbb{Z}$ and $(\alpha)^* = 0$ otherwise, and

- $\varepsilon(m)$ is a term that depends only on the divisors of m which are less than 180.

It is known that $\varepsilon(m) = 0$ if $\gcd(m, 6) = 1$; otherwise, its value has to be determined from the table in [Sh82a]. Aoki and Shioda [AS83] have described quite precisely the algebraic cycles corresponding to all but the last term in this formula.

(b) We now compute the Picard numbers of diagonal hypersurfaces $\mathcal{V} = \mathcal{V}_2^m(\mathbf{c})$ for selected twists $\mathbf{c}$. For each prime m, let p be a prime such that $p \equiv 1 \pmod{m}$. Choose a primitive root g modulo p. We will consider twists $\mathbf{c} = (c_0, c_1, c_2, c_3)$ where each component c_i is of the form g^j with $1 \leq j \leq m-1$. (This avoids duplication, since we know that twists whose entries differ by m-th powers will be isomorphic.) We observe from our computations the following facts:

1. All twists of the form $(g^j, 1, 1, 1)$ with $1 \leq j \leq m-1$ are extreme, i.e., their Picard number is 1. As pointed out in the main text, this is clear, since in this case $\mathbf{c^a} = g^{ja_0}$, which cannot be an m-th power, since g is a primitive root and a_0 is not a multiple of m.

2. For any prime m, all twists of the form $(g^j, g^j, 1, 1)$ for $1 \leq j \leq m-1$ seem to have the same Picard number, and in fact it is always equal to $1 + (m-1)^2$. We offer no explanation for this fact.

(These two types of twists will in general not be listed in the tables.)

(b1) Let $m = 5$, $n = 2$ and $p = 11$. We compute the Picard number of $\mathcal{V}(\mathbf{c})$ for various twists $\mathbf{c}$. Recall that the stable Picard number is 37.

c	$\rho_1(\mathcal{V}_k)$	c	$\rho_1(\mathcal{V}_k)$	c	$\rho_1(\mathcal{V}_k)$
$(4,2,1,1)$	9	$(8,8,2,1)$	5	$(5,8,1,1)$	9
$(4,2,2,1)$	5	$(8,8,4,1)$	9	$(5,8,2,1)$	9
$(4,4,2,1)$	9	$(5,2,1,1)$	5	$(5,8,4,1)$	9
$(8,2,1,1)$	9	$(5,2,2,1)$	9	$(5,5,2,1)$	9
$(8,4,1,1)$	5	$(5,4,1,1)$	9	$(5,5,4,1)$	9
$(8,4,2,1)$	9	$(5,4,2,1)$	9	$(5,5,8,1)$	5
$(8,4,4,1)$	9	$(5,4,4,1)$	5	$(2,1,1,1)$	1

(b2) Let $m = 7$, $n = 2$ and $p = 29$ (which is congruent to 1 modulo 7). We compute the Picard number of $\mathcal{V}(\mathbf{c})$ for various twists. In this case the stable Picard number is 91.

c	$\rho_1(\mathcal{V}_k)$	c	$\rho_1(\mathcal{V}_k)$	c	$\rho_1(\mathcal{V}_k)$
$(4,2,1,1)$	13	$(3,2,1,1)$	13	$(6,8,4,1)$	13
$(8,2,1,1)$	13	$(3,4,1,1)$	7	$(6,16,2,1)$	19
$(8,4,1,1)$	13	$(3,4,2,1)$	19	$(6,3,2,1)$	13
$(8,4,2,1)$	13	$(3,16,4,1)$	13	$(6,3,4,1)$	19
$(4,4,2,1)$	13	$(3,16,8,1)$	19	$(6,3,8,1)$	19
$(16,4,2,1)$	19	$(6,2,1,1)$	7	$(6,3,16,1)$	13
$(16,8,1,1)$	7	$(6,4,2,1)$	13	$(6,6,2,1)$	13
$(16,8,2,1)$	13	$(6,8,1,1)$	13	$(6,6,4,1)$	13
$(16,16,2,1)$	7	$(6,8,2,1)$	19	$(6,6,3,1)$	7

(b3) Let $m = 17$, $n = 2$ and $p = 103$. The stable Picard number is 721.

c	$\rho_1(\mathcal{V}_k)$	c	$\rho_1(\mathcal{V}_k)$	c	$\rho_1(\mathcal{V}_k)$
$(25,5,1,1)$	33	$(39,35,7,1)$	33	$(26,35,7,1)$	49
$(22,5,1,1)$	33	$(39,49,1,1)$	17	$(26,48,49,1)$	33
$(7,25,5,1)$	49	$(92,72,7,11)$	33	$(26,67,35,1)$	49
$(7,22,5,1)$	33	$(92,51,1,1)$	17	$(27,25,1,1)$	17

c	$\rho_1(\mathcal{V}_k)$	c	$\rho_1(\mathcal{V}_k)$	c	$\rho_1(\mathcal{V}_k)$
$(35, 22, 25, 1)$	49	$(92, 39, 51, 1)$	49	$(27, 51, 22, 1)$	49
$(35, 5, 1, 1)$	33	$(48, 72, 1, 1)$	17	$(27, 92, 35, 1)$	33
$(72, 7, 5, 1)$	49	$(48, 38, 49, 1)$	49	$(27, 26, 67, 1)$	49
$(72, 35, 5, 1)$	33	$(34, 35, 1, 1)$	17	$(32, 5, 1, 1)$	17
$(51, 35, 25, 1)$	33	$(34, 48, 92, 1)$	49	$(32, 51, 7, 1)$	49
$(51, 72, 7, 1)$	49	$(67, 7, 1, 1)$	17	$(32, 48, 35, 1)$	33
$(49, 72, 5, 1)$	49	$(67, 39, 35, 1)$	33	$(32, 92, 49, 1)$	49
$(49, 35, 22, 1)$	33	$(26, 22, 1, 1)$	17	$(32, 27, 26, 1)$	33
$(49, 5, 1, 1)$	33	$(26, 5, 1, 1)$	33	$(32, 7, 1, 1)$	33
$(32, 22, 22, 1)$	33	$(32, 27, 48, 1)$	49	$(32, 32, 1, 1)$	257
$(32, 32, 27, 1)$	17	$(32, 32, 26, 1)$	33	$(32, 35, 1, 1)$	33
$(25, 5, 5, 1)$	17	$(25, 25, 5, 1)$	33	$(22, 25, 1, 1)$	33

(b4) Let $m = 29$, $n = 2$ and $p = 59$. The stable Picard number is 2269.

c	$\rho_1(\mathcal{V}_k)$	c	$\rho_1(\mathcal{V}_k)$	c	$\rho_1(\mathcal{V}_k)$
$(4, 2, 1, 1)$	57	$(46, 50, 1, 1)$	29	$(47, 7, 33, 1)$	85
$(8, 2, 1, 1)$	57	$(46, 23, 2, 1)$	57	$(47, 28, 41, 1)$	57
$(8, 4, 2, 1)$	57	$(46, 23, 41, 1)$	85	$(35, 32, 1, 1)$	29
$(16, 4, 2, 1)$	85	$(33, 21, 10, 1)$	57	$(35, 7, 1, 1)$	57
$(16, 8, 2, 1)$	57	$(33, 25, 1, 1)$	29	$(35, 14, 32, 1)$	57
$(32, 2, 1, 1)$	57	$(33, 41, 25, 1)$	85	$(35, 53, 16, 1)$	85
$(32, 16, 8, 1)$	85	$(7, 8, 1, 1)$	57	$(35, 47, 53, 1)$	85
$(5, 16, 4, 1)$	57	$(7, 42, 1, 1)$	29	$(11, 16, 1, 1)$	29
$(5, 32, 16, 1)$	85	$(7, 23, 10, 1)$	85	$(11, 5, 1, 1)$	57
$(10, 8, 4, 1)$	85	$(14, 21, 1, 1)$	29	$(11, 42, 10, 1)$	57
$(10, 5, 2, 1)$	57	$(14, 25, 4, 1)$	57	$(11, 7, 10, 1)$	57
$(20, 16, 4, 1)$	85	$(14, 7, 33, 1)$	85	$(11, 35, 47, 1)$	85
$(20, 10, 2, 1)$	57	$(28, 40, 1, 1)$	29	$(22, 8, 1, 1)$	29
$(40, 5, 16, 1)$	85	$(28, 41, 21, 1)$	85	$(22, 7, 23, 1)$	57

c	$\rho_1(\mathcal{V}_k)$	c	$\rho_1(\mathcal{V}_k)$	c	$\rho_1(\mathcal{V}_k)$
$(40, 20, 2, 1)$	57	$(28, 7, 4, 1)$	57	$(22, 53, 1, 1)$	57
$(21, 5, 16, 1)$	57	$(56, 20, 1, 1)$	29	$(22, 11, 2, 1)$	57
$(21, 20, 5, 1)$	85	$(56, 40, 2, 1)$	57	$(22, 11, 35, 1)$	85
$(42, 5, 1, 1)$	57	$(56, 25, 16, 1)$	57	$(44, 4, 1, 1)$	29
$(42, 21, 5, 1)$	85	$(56, 7, 1, 1)$	57	$(44, 41, 25, 1)$	57
$(25, 5, 16, 1)$	85	$(56, 28, 5, 1)$	85	$(44, 56, 5, 1)$	57
$(25, 42, 2, 1)$	57	$(53, 5, 1, 1)$	57	$(44, 47, 1, 1)$	57
$(50, 20, 2, 1)$	85	$(53, 10, 1, 1)$	29	$(44, 35, 53, 1)$	57
$(50, 42, 4, 1)$	57	$(53, 21, 8, 1)$	57	$(44, 22, 11, 1)$	85
$(41, 40, 8, 1)$	85	$(53, 7, 16, 1)$	57	$(29, 2, 1, 1)$	29
$(41, 50, 2, 1)$	57	$(53, 56, 28, 1)$	85	$(29, 7, 5, 1)$	85
$(23, 40, 5, 1)$	57	$(47, 5, 1, 1)$	29	$(29, 14, 40, 1)$	57
$(23, 41, 1, 1)$	29	$(47, 25, 1, 1)$	57	$(29, 44, 1, 1)$	57
$(23, 41, 50, 1)$	85	$(47, 46, 21, 1)$	57	$(29, 44, 22, 1)$	57

Notice that in all the examples the Picard number of $\mathcal{V}(\mathbf{c})$ is substantially smaller than the stable Picard number, which (for m prime) is also the Picard number of the (untwisted) Fermat hypersurface $\mathcal{X}$. This pattern changes markedly when m is not prime.

(c) Finally, we compute the actual Picard numbers of diagonal hypersurfaces $\mathcal{V}_k$ of dimension $n = 2$ and degree m where m is *composite*. We choose $k = \mathbb{F}_q$ with $q = p \equiv 1 \pmod{m}$, and pick a primitive root g modulo p. We will consider twists $\mathbf{c} = (c_0, c_1, c_2, c_3)$ such that each component c_i is of the form g^j, $1 \le j \le m - 1$.

If m is an odd prime power, the Iwasawa congruence still holds, and hence so does Proposition 1.10. This places some constraints on the Picard numbers. For more general m, the Iwasawa congruence does not hold, and things become quite complex, which makes these computations particularly interesting.

We observed certain patterns in the results of our computations:

1. When $m = m_0^r$ where m_0 is an odd prime ≥ 3 and $r \ge 2$, some (but not all) twists of the form $(g^j, 1, 1, 1)$ are extreme. For instance, if $m = 9$ or 25, then $\mathbf{c}$ is extreme when $j \equiv 2 \pmod 3$, $1 \le j \le m - 1$. The case of $m = 49$ proved to be beyond the reach of our computations.

2. When $m = m_0 \cdot m_1$ where m_0 and m_1 are (distinct) odd primes, then some (but not all) twists of the form $\mathbf{c} = (g^j, 1, 1, 1)$ are extreme. There seems to be no clear pattern in this case, and there are very few extreme twists in any case.

3. In both cases (1) and (2), the equality $\rho_1(\mathcal{V}_k) \leq \rho_1(\mathcal{X}_k)$ holds in all the examples we computed. We have indicated this below, and have highlighted in bold those cases when the inequality fails. Notice also that even when the inequality does hold, it does not hold by as wide a margin as in the case of prime degree.

4. When m is even, there are twists $\mathbf{c}$ for which $\rho_1(\mathcal{V}_k) \geq \rho_1(\mathcal{X}_k)$. We are not able to detect any pattern on the actual Picard numbers. The situation seems rather wild especially when $(m, 6) > 1$.

5. We notice that even though for composite m one can have $\rho_1(\mathcal{V}_k) > \rho_1(\mathcal{X}_k)$, the value of the Picard number is always substantially smaller than the stable Picard number.

Here are some computational results.

(c1) Let $(m, n) = (4, 2)$, $p = 5$ and take $g = 2$. The stable Picard number is 20. The inequality $\rho_1(\mathcal{V}_k) \leq \rho_1(\mathcal{X}_k)$ does not hold; furthermore, one can have $\rho_1(\mathcal{V}_k) = \rho_1(\mathcal{X}_k)$ without having $\mathcal{V}_k \cong \mathcal{X}_k$. Examples of both phenomena are in bold.

$\mathbf{c}$	$\rho_1(\mathcal{V}_k)$	$\mathbf{c}$	$\rho_1(\mathcal{V})$	$\mathbf{c}$	$\rho_1(\mathcal{V}_k)$
$(1, 1, 1, 1)$	8	$(3, 2, 1, 1)$	6	$(3, 3, 2, 1)$	3
$(2, 2, 1, 1)$	6	$(3, 2, 2, 1)$	3	$(3, 3, 4, 1)$	6
$(4, 2, 1, 1)$	3	$(3, 4, 1, 1)$	3	$(2, 1, 1, 1)$	7
$(4, 2, 2, 1)$	6	$(3, 4, 2, 1)$	**10**	$(4, 1, 1, 1)$	**8**
$(4, 4, 1, 1)$	**16**	$(3, 4, 4, 1)$	3	$(3, 1, 1, 1)$	7
$(4, 4, 2, 1)$	3	$(3, 3, 1, 1)$	6	$(2, 3, 1, 1)$	3

(c2) Let $(m, n) = (9, 2)$, $p = 19$ and take $g = 2$. The stable Picard number is 217. The inequality $\rho_1(\mathcal{V}_k) \leq \rho_1(\mathcal{X}_k)$ holds.

c	$\rho_1(\mathcal{V}_k)$	c	$\rho_1(\mathcal{V}_k)$	c	$\rho_1(\mathcal{V}_k)$
$(1,1,1,1)$	169	$(8,4,1,1)$	25	$(7,7,7,1)$	43
$(2,2,1,1)$	65	$(8,8,1,1)$	115	$(14,14,14,1)$	1
$(2,2,2,1)$	1	$(8,8,8,1)$	79	$(14,7,13,1)$	33
$(4,2,1,1)$	27	$(16,4,2,1)$	33	$(9,9,2,1))$	27
$(4,2,2,1)$	21	$(16,16,16,1)$	1	$(9,9,14,1)$	21
$(4,4,2,1)$	15	$(7,8,1,1)$	61	$(4,1,1,1)$	1
$(4,4,4,1)$	13	$(7,8,2,1)$	31	$(13,1,1,1)$	1
$(8,2,2,1)$	29	$(7,8,4,1)$	19	$(9,1,1,1)$	1

(c3) Let $(m,n) = (12,2)$, $p = 13$ and take $g = 2$. The stable Picard number is 644. The inequality $\rho_1(\mathcal{V}_k) \leq \rho_1(\mathcal{X}_k)$ does not hold.

c	$\rho_1(\mathcal{V}_k)$	c	$\rho_1(\mathcal{V}_k)$	c	$\rho_1(\mathcal{V}_k)$
$(1,1,1,1)$	152	$(3,3,8,1)$	93	$(12,12,12,1)$	**188**
$(2,2,1,1)$	106	$(3,3,3,1)$	44	$(11,4,2,1)$	80
$(2,2,2,1)$	43	$(6,2,1,1)$	60	$(11,8,8,1)$	51
$(4,2,1,1)$	37	$(6,2,2,1)$	69	$(11,3,1,1)$	113
$(4,2,2,1)$	48	$(6,4,1,1)$	57	$(11,3,3,1)$	47
$(4,4,1,1)$	**164**	$(6,8,1,1)$	38	$(11,6,2,1)$	45
$(4,4,2,1)$	49	$(6,8,4,1)$	118	$(11,6,3,1)$	46
$(4,4,4,1)$	98	$(6,8,8,1)$	27	$(11,12,2,1)$	**174**
$(8,2,1,1)$	76	$(6,3,2,1)$	46	$(11,12,4,1)$	29
$(8,2,2,1)$	89	$(6,3,8,1)$	42	$(11,12,3,1)$	81
$(8,4,1,1)$	35	$(6,3,3,1)$	33	$(11,11,8,1)$	61
$(8,4,2,1)$	70	$(6,6,1,1)$	106	$(11,11,3,1)$	66
$(8,4,4,1)$	77	$(6,6,6,1)$	91	$(11,11,11,1)$	19
$(8,8,1,1)$	**154**	$(12,2,2,1)$	110	$(9,4,1,1)$	128
$(8,8,2,1)$	47	$(12,4,4,1)$	120	$(9,3,1,1)$	50
$(8,8,4,1)$	54	$(12,8,1,1)$	107	$(9,12,4,1)$	**196**
$(8,8,8,1)$	79	$(12,8,2,1)$	116	$(9,12,8,1)$	31
$(3,2,1,1)$	65	$(12,8,8,1)$	134	$(9,11,1,1)$	49
$(3,2,2,1)$	50	$(12,3,4,1)$	102	$(9,11,3,1)$	51

c	$\rho_1(\mathcal{V}_k)$	c	$\rho_1(\mathcal{V}_k)$	c	$\rho_1(\mathcal{V}_k)$
$(3,4,1,1)$	62	$(12,3,8,1)$	71	$(9,9,8,1)$	113
$(3,4,2,1)$	35	$(12,3,3,1)$	128	$(9,9,9,1)$	62
$(3,8,1,1)$	115	$(12,6,1,1)$	15	$(5,8,1,1)$	134
$(3,8,2,1)$	82	$(12,6,2,1)$	48	$(5,8,4,1)$	74
$(3,8,4,1)$	39	$(12,6,4,1)$	69	$(5,12,8,1)$	**222**
$(3,8,8,1)$	96	$(12,6,8,1)$	74	$(5,5,5,1)$	127
$(3,3,1,1)$	92	$(12,12,1,1)$	**288**	$(10,10,10,1)$	56
$(3,3,2,1)$	23	$(12,12,2,1)$	55	$(7,5,5,1)$	34
$(3,1,1,1)$	61	$(12,1,1,1)$	**187**	$(7,7,1,1)$	105

(c4) Let $(m,n) = (15,2)$, $p = 31$ and take $g = 3$. The stable Picard number is 835. The inequality $\rho_1(\mathcal{V}_k) \leq \rho_1(\mathcal{X}_k)$ holds by a wide margin.

c	$\rho_1(\mathcal{V}_k)$	c	$\rho_1(\mathcal{V})$	c	$\rho_1(\mathcal{V}_k)$
$(1,1,1,1)$	547	$(19,9,9,1)$	23	$(16,16,1,1)$	295
$(3,3,1,1)$	213	$(19,19,1,1)$	213	$(16,16,27,1)$	103
$(3,3,3,1)$	1	$(26,3,3,1)$	91	$(16,16,16,1)$	79
$(9,3,1,1)$	27	$(26,9,1,1)$	69	$(17,26,27,1)$	39
$(9,3,3,1)$	15	$(26,9,9,1)$	73	$(17,16,9,1)$	51
$(9,9,1,1)$	197	$(26,27,1,1)$	17	$(17,17,1,1)$	197
$(20,3,1,1)$	35	$(20,3,3,1)$	27	$(20,9,1,1)$	57
$(20,9,3,1)$	43	$(20,9,9,1)$	49	$(20,27,1,1)$	73
$(20,27,3,1)$	31	$(20,27,9,1)$	61	$(20,27,27,1)$	57
$(20,19,1,1)$	27	$(20,19,3,1)$	35	$(20,19,9,1)$	35
$(20,19,27,1)$	35	$(20,19,19,1)$	15	$(20,26,1,1)$	45
$(20,26,3,1)$	35	$(20,26,9,1)$	37	$(20,26,27,1)$	101
$(20,26,19,1)$	43	$(20,26,26,1)$	17	$(20,16,1,1)$	49
$(20,16,3,1)$	35	$(20,16,9,1)$	61	$(20,16,27,1)$	37
$(20,16,19,1)$	43	$(20,16,26,1)$	69	$(20,16,16,1)$	25
$(20,17,1,1)$	23	$(20,17,3,1)$	27	$(20,17,9,1)$	43
$(20,17,27,1)$	35	$(20,17,19,1)$	43	$(20,17,26,1)$	39
$(20,17,16,1)$	35	$(20,17,17,1)$	35	$(20,20,1,1)$	197

c	$\rho_1(\mathcal{V}_k)$	c	$\rho_1(\mathcal{V})$	c	$\rho_1(\mathcal{V}_k)$
$(20,20,3,1)$	23	$(20,20,9,1)$	25	$(20,20,27,1)$	69
$(20,20,19,1)$	27	$(20,20,26,1)$	73	$(20,20,16,1)$	57
$(27,3,1,1)$	33	$(26,27,9,1)$	57	$(20,20,17,1)$	27
$(27,3,3,1)$	53	$(26,19,1,1)$	19	$(26,26,19,1)$	19
$(27,9,1,1)$	45	$(26,19,19,1)$	75	$(29,27,27,1)$	103
$(27,9,3,1)$	35	$(26,26,1,1)$	297	$(29,26,27,1)$	65
$(27,9,9,1)$	61	$(26,26,26,1)$	157	$(25,26,1,1)$	167
$(27,27,1,1)$	247	$(16,3,1,1)$	89	$(25,16,3,1)$	117
$(27,27,3,1)$	29	$(16,9,3,1)$	31	$(3,1,1,1)$	25
$(27,27,9,1)$	25	$(16,27,1,1)$	127	$(27,1,1,1)$	79
$(27,27,27,1)$	151	$(16,27,27,1)$	115	$(19,1,1,1)$	1
$(19,3,1,1)$	49	$(16,19,1,1)$	21	$(26,1,1,1)$	157
$(19,3,3,1)$	41	$(16,19,3,1)$	77	$(16,1,1,1)$	103
$(19,9,3,1)$	43	$(16,26,3,1)$	83	$(20,1,1,1)$	1

A.3.2 Dimension 4. (a) We compute the Picard numbers and the stable Picard numbers of diagonal hypersurfaces with trivial twist of dimension $n = 4$ and degree m, where $4 \leq m \leq 47$. As before, we simplify the computational task by choosing $k = \mathbb{F}_q$ with $q = p \equiv 1 \pmod{m}$, so that all our motives are ordinary and the Tate conjecture is known to hold when m is prime and when $m \leq 21$.

m	p	$\rho_2(\mathcal{X}_k)$	$\bar{\rho}_2(\mathcal{V})$	m	p	$\rho_2(\mathcal{X}_k)$	$\bar{\rho}_2(\mathcal{V})$
4	5	92	142	26	53		439082
5	11	401	401	27	109		301941
6	13	591	1752	28	29	139412	644122
7	29	1861	1861	29	59	295121	295121
8	17	3482	5882	30	31	209811	2106432
9	19	5121	8001	31	311	365701	365701
10	11	4061	19882	32	97		727202
11	23	10901	10901	33	67		5557601
12	13	10152	52992	34	103		909922
13	43	19921	19921	35	71		543221
14	29	32942	77402	36	73		1923972
15	31	32901	78261	37	149	642961	642961
16	17	29762	87992	38	191		1241822
17	103	50561	50561	39	79		923781
18	37	34041	264672	40	41		2016682
19	191	73621	73621	41	83	889601	889601
20	41	90542	346382	42	43		5679132
21	43	102801	215121	43	173	1033621	1033621
22	23	60551	277102	44	89		1850792
23	47	138821	138821	45	271		1543041
24	73	159222	745212	46	47		2207482
25	51	1946	185281	47	283		

For m prime, our computational results are consistent with Proposition 5.4:

$$\bar{\rho}_2(\mathcal{V}) = 1 + 5(m-1)(3m^2 - 15m + 20).$$

We first obtained this formula simply by assuming that the stable Picard numbers should be given by a cubic polynomial in m, and then determined that polynomial by considering the results for $m = 5, 7, 11,$ and 13. We then checked that the answers given by this formula match those we obtained for all primes up to 47. A proof that the formula does indeed hold for all prime degrees can be given using the work of Shioda, and is described in Appendix B.

(b) We compute the Picard numbers of the diagonal hypersurfaces $\mathcal{V} = \mathcal{V}_4^m(\mathbf{c})$ for selected twists $\mathbf{c}$. Fix m prime, and as before simplify the computation by picking a prime p such that $p \equiv 1 \pmod{m}$. Choose a primitive root

g modulo p. We will be considering twisting vectors $\mathbf{c} = (c_0, c_1, c_2, c_3, c_4, c_5)$ where each c_i is of the form g^j with $1 \le j \le m - 1$. (This avoids considering twists that are known to produce isomorphic diagonal hypersurfaces.)

(b1) Let $m = 5$, $n = 4$ and $p = 11$; the stable Picard number is 401.

$\mathbf{c}$	$\rho_2(\mathcal{V}_k)$	$\mathbf{c}$	$\rho_2(\mathcal{V}_k)$	$\mathbf{c}$	$\rho_2(\mathcal{V}_k)$
$(2,2,1,1,1,1)$	145	$(8,8,2,1,1,1)$	61	$(5,8,2,1,1,1)$	85
$(2,2,2,1,1,1)$	37	$(8,8,4,4,2,1)$	77	$(5,8,4,2,1,1)$	81
$(4,2,1,1,1,1)$	97	$(5,2,1,1,1,1)$	65	$(5,8,8,2,1,1)$	77
$(4,2,2,1,1,1)$	61	$(5,2,2,1,1,1)$	85	$(5,5,2,1,1,1)$	85
$(8,2,1,1,1,1)$	97	$(5,4,1,1,1,1)$	97	$(5,5,4,4,1,1)$	97
$(8,4,1,1,1,1)$	65	$(5,4,2,1,1,1)$	85	$(5,5,8,1,1,1)$	61
$(8,4,2,1,1,1)$	85	$(5,4,4,1,1,1)$	61	$(5,8,4,2,1,1)$	81
$(8,4,4,2,1,1)$	77	$(5,8,1,1,1,1)$	97	$(5,5,8,8,8,1)$	61

As in the case of dimension 2, we note that the Picard numbers of the twisted hypersurfaces are substantially smaller than the stable Picard number (which is equal to the Picard number of the Fermat hypersurface). This will continue to be true in all the examples of prime degree.

(b2) Let $m = 7$, $n = 4$ and $p = 29$, and take $g = 2$. In this case the stable Picard number is 1861.

$\mathbf{c}$	$\rho_2(\mathcal{V}_k)$	$\mathbf{c}$	$\rho_2(\mathcal{V}_k)$	$\mathbf{c}$	$\rho_2(\mathcal{V}_k)$
$(2,2,1,1,1,1)$	541	$(3,4,1,1,1,1)$	169	$(6,3,4,2,1,1)$	265
$(2,2,2,1,1,1)$	127	$(3,3,2,1,1,1)$	235	$(6,3,8,4,2,1)$	259
$(4,2,1,1,1,1)$	289	$(3,8,2,1,1,1)$	253	$(6,3,16,2,1,1)$	283
$(4,2,2,1,1,1)$	199	$(3,16,8,1,1,1)$	307	$(6,3,3,3,4,1)$	313
$(4,4,2,2,1,1)$	361	$(3,16,4,2,1,1)$	259	$(6,3,16,8,4,1)$	259
$(8,4,2,1,1,1)$	253	$(3,16,4,2,2,1)$	265	$(6,18,8,1,1,1)$	271
$(16,8,4,1,1,1)$	307	$(3,3,8,8,1,1)$	361	$(6,6,4,4,1,1)$	337
$(16,8,8,1,1,1)$	271	$(6,8,4,2,1,1)$	283	$(6,16,2,1,1,1)$	307
$(16,8,8,2,2,1)$	247	$(6,3,3,3,1,1)$	235	$(6,6,3,1,1,1)$	199

(b3) Let $m = 13$, $n = 4$ and $p = 53$, and take $g = 2$. In this case the stable Picard number is 19921.

c	$\rho_2(\mathcal{V}_k)$	c	$\rho_2(\mathcal{V}_k)$
$(2,2,1,1,1,1)$	4753	$(16,8,4,4,2,1)$	1537
$(2,2,2,1,1,1)$	685	$(32,8,4,2,1,1)$	1561
$(4,2,1,1,1,1)$	1441	$(32,8,4,4,2,1)$	1573
$(4,2,2,1,1,1)$	1045	$(32,8,8,4,4,1)$	1621
$(4,4,2,1,1,1)$	1333	$(32,16,4,2,1,1)$	1489
$(8,4,2,1,1,1)$	1369	$(32,16,8,2,2,1)$	1609
$(16,4,2,1,1,1)$	1657	$(32,16,8,8,8,1)$	1705
$(32,16,16,1,1,1)$	1381	$(11,8,4,2,1,1)$	1585
$(11,16,8,1,1,1)$	1705	$(11,11,8,4,2,1)$	1465
$(22,11,1,1,1,1)$	769	$(22,11,2,1,1,1)$	1405
$(22,11,4,1,1,1)$	1693	$(4,4,2,2,1,1)$	2593
$(8,4,2,2,1,1)$	1525	$(8,4,4,2,1,1)$	1453
$(8,4,4,2,2,1)$	1549	$(8,8,4,4,1,1)$	2545
$(8,8,4,2,1,1)$	1501	$(16,4,4,2,1,1)$	1477
$(16,4,4,2,2,1)$	1429	$(16,8,4,2,1,1)$	1513
$(16,8,4,2,2,1)$	1597	$(16,2,2,2,1,1)$	1333
$(16,2,2,2,2,1)$	1441	$(16,4,4,4,4,1)$	769
$(16,16,8,8,4,1)$	1525	$(16,16,16,4,4,1)$	1045
$(16,16,16,16,1,1)$	4753	$(16,16,16,16,16,1)$	1
$(16,16,2,2,1,1)$	2545	$(16,16,2,2,2,1)$	1333

(b4) Let $m = 17$, $n = 4$ and $p = 103$, and take $g = 5$. The stable Picard number is 50561.

c	$\rho_2(\mathcal{V}_k)$	c	$\rho_2(\mathcal{V}_k)$
$(5,5,1,1,1,1)$	11521	$(34,35,7,5,1,1)$	3025
$(5,5,5,1,1,1)$	1297	$(34,49,51,72,5,1)$	2945
$(25,5,1,1,1,1)$	2689	$(67,51,35,25,5,1)$	2993

c	$\rho_2(\mathcal{V}_k)$	c	$\rho_2(\mathcal{V}_k)$
$(25, 5, 5, 1, 1, 1)$	1969	$(67, 49, 51, 51, 51, 1)$	3233
$(25, 25, 5, 5, 1, 1)$	5761	$(67, 67, 7, 7, 7, 1)$	2545
$(22, 25, 5, 5, 1, 1)$	2929	$(26, 7, 5, 5, 5, 1)$	2593
$(22, 22, 5, 5, 1, 1)$	5697	$(26, 35, 7, 25, 1, 1)$	3073
$(22, 22, 25, 25, 25, 1)$	2609	$(26, 26, 22, 22, 1, 1)$	5761
$(7, 25, 5, 1, 1, 1)$	3169	$(27, 7, 25, 25, 25, 1)$	2641
$(7, 25, 25, 5, 1, 1)$	2865	$(27, 35, 7, 22, 5, 1)$	2945
$(7, 22, 25, 5, 5, 1)$	3089	$(27, 72, 25, 5, 5, 1)$	3057
$(35, 22, 25, 5, 1, 1)$	3041	$(32, 51, 72, 22, 5, 1)$	2977
$(35, 7, 22, 5, 5, 1)$	3105	$(32, 49, 5, 5, 5, 1)$	3169
$(72, 22, 25, 5, 5, 1)$	3025	$(32, 49, 7, 5, 1, 1)$	2993
$(72, 7, 22, 22, 25, 1)$	2881	$(51, 22, 25, 5, 1, 1)$	3009
$(51, 7, 7, 5, 1, 1)$	2705	$(51, 35, 22, 22, 1, 1)$	2833
$(49, 25, 5, 5, 5, 1)$	3217	$(49, 7, 25, 5, 1, 1)$	2977
$(49, 72, 35, 5, 5, 1)$	3073	$(39, 7, 22, 22, 22, 1)$	3169
$(39, 72, 22, 25, 5, 1)$	2913	$(92, 5, 1, 1, 1, 1)$	1409
$(92, 35, 22, 22, 25, 1)$	3009	$(92, 51, 7, 25, 5, 1)$	2961
$(48, 35, 7, 22, 5, 1)$	2977	$(48, 92, 39, 72, 22, 1)$	2913

(c) We compute the actual Picard numbers of diagonal hypersurfaces $\mathcal{V}_k$ of dimension $n = 4$ and degree m where m is *composite*. We choose $k = \mathbb{F}_q$ with $q = p \equiv 1 \pmod{m}$, and pick a primitive root g modulo p. We will consider twists $\mathbf{c} = (c_0, c_1, c_2, c_3, c_4, c_5)$ such that each component c_i is of the form g^j, $1 \le j \le m - 1$. Here we observe the following facts from our computations:

1. For any composite $m \in \{4, 6, 8, 9, 10\}$, there were no extreme twists.

2. When $m = 9$, the inequality $\rho_2(\mathcal{V}_k) \le \rho_2(\mathcal{X}_k)$ holds.

3. When $m \in \{4, 6, 8, 10\}$, there are twists $\mathbf{c}$ for which $\rho_2(\mathcal{V}_k) \ge \rho_2(\mathcal{X}_k)$.

We were not able to detect any pattern in the actual Picard numbers. Here are some computational results.

(c1) Let $(m, n) = (6, 4)$, $p = 7$ and take $g = 3$. The stable Picard number is 1752. The inequality $\rho_1(\mathcal{V}_k) \le \rho_1(\mathcal{X}_k)$ does not hold.

c	$\rho_2(\mathcal{V}_k)$	c	$\rho_2(\mathcal{V})$	c	$\rho_2(\mathcal{V}_k)$
$(1,1,1,1,1,1)$	591	$(4,6,6,6,6,1)$	177	$(6,2,1,1,1,1)$	258
$(3,3,3,3,1,1)$	343	$(4,4,6,6,6,1)$	448	$(6,6,2,1,1,1)$	309
$(3,3,3,3,3,1)$	282	$(4,4,4,2,3,1)$	208	$(6,6,6,1,1,1)$	**666**
$(2,3,3,3,3,1)$	229	$(5,2,3,3,3,1)$	273	$(4,3,1,1,1,1)$	318
$(2,2,3,3,3,1)$	394	$(5,2,2,2,2,1)$	164	$(4,4,2,1,1,1)$	357
$(2,2,2,3,1,1)$	230	$(5,6,2,2,2,1)$	329	$(5,4,2,1,1,1)$	216
$(2,2,2,2,1,1)$	447	$(5,6,6,2,1,1)$	472	$(5,5,2,1,1,1)$	227
$(2,2,2,2,3,1)$	190	$(5,4,3,3,3,1)$	235	$(5,5,6,1,1,1)$	430
$(6,3,3,3,1,1)$	227	$(5,4,6,6,2,1)$	328	$(5,5,4,1,1,1)$	231
$(6,2,3,3,1,1)$	382	$(5,4,4,4,3,1)$	259	$(5,5,5,1,1,1)$	342
$(6,2,2,3,3,1)$	414	$(5,5,5,5,1,1)$	351	$(2,1,1,1,1,1)$	441
$(6,6,6,3,1,1)$	133	$(5,5,5,5,5,1)$	92	$(6,1,1,1,1,1)$	442
$(4,2,2,2,3,1)$	216	$(3,3,3,1,1,1)$	342	$(4,1,1,1,1,1)$	201
$(4,6,6,3,1,1)$	502	$(2,2,1,1,1,1)$	423	$(5,1,1,1,1,1)$	282

(c2) Let $(m,n) = (9,4)$, $p = 19$ and take $g = 2$. The stable Picard number is 8001. The inequality $\rho_1(\mathcal{V}_k) \leq \rho_1(\mathcal{X}_k)$ holds by a wide margin.

c	$\rho_2(\mathcal{V}_k)$	c	$\rho_2(\mathcal{V}_k)$
$(1,1,1,1,1,1)$	5121	$(7,7,7,7,1,1)$	3441
$(2,2,2,2,1,1)$	1537	$(14,2,2,2,2,1)$	553
$(2,2,2,2,2,1)$	121	$(14,7,13,16,8,1)$	877
$(4,4,2,2,1,1)$	1089	$(14,14,7,16,2,1)$	1009
$(4,4,4,4,4,1)$	601	$(9,4,4,4,1,1)$	1039
$(8,4,2,2,2,1)$	841	$(9,14,7,16,1,1)$	901
$(8,8,8,8,1,1)$	3297	$(2,2,1,1,1,1)$	1489
$(8,8,8,8,2,1)$	529	$(4,2,2,1,1,1)$	781
$(16,4,4,4,1,1)$	745	$(4,4,1,1,1,1)$	1537
$(16,8,4,2,2,1)$	883	$(7,1,1,1,1,1)$	3081
$(16,16,8,8,1,1)$	1225	$(9,1,1,1,1,1)$	121

c	$\rho_2(\mathcal{V}_k)$	c	$\rho_2(\mathcal{V}_k)$
$(13, 8, 4, 4, 1, 1)$	813	$(13, 16, 8, 4, 2, 1)$	891
$(13, 13, 4, 4, 4, 1)$	619	$(13, 13, 13, 13, 1, 1)$	1489
$(7, 8, 8, 8, 1, 1)$	2451	$(7, 13, 4, 4, 1, 1)$	777
$(7, 13, 13, 13, 16, 1)$	907	$(7, 7, 8, 8, 8, 1)$	2631
$(7, 7, 13, 16, 4, 1)$	925	$(7, 7, 13, 13, 1, 1)$	1201
$(7, 7, 7, 4, 1, 1)$	535	$(7, 1, 1, 1, 1, 1)$	3080

(c3) Let $(m, n) = (10, 4)$, $p = 11$ and take $g = 2$. The stable Picard number is 19882. The inequality $\rho_1(\mathcal{V}_k) \leq \rho_1(\mathcal{X}_k)$ does not hold.

c	$\rho_1(\mathcal{V}_k)$	c	$\rho_1(\mathcal{V}_k)$
$(1, 1, 1, 1, 1, 1)$	4061	$(10, 10, 4, 1, 1, 1)$	1057
$(2, 2, 2, 2, 1, 1)$	2221	$(10, 10, 10, 1, 1, 1)$	**5218**
$(4, 4, 4, 4, 1, 1)$	3053	$(9, 10, 1, 1, 1, 1)$	1258
$(8, 8, 8, 8, 8, 1)$	722	$(7, 5, 1, 1, 1, 1)$	1082
$(5, 5, 8, 4, 1, 1)$	1510	$(3, 3, 7, 1, 1, 1)$	1178
$(5, 5, 5, 2, 1, 1)$	1462	$(6, 7, 2, 1, 1, 1)$	2474
$(10, 5, 4, 4, 4, 1)$	1202	$(2, 1, 1, 1, 1, 1)$	1322
$(9, 8, 8, 2, 1, 1)$	2818	$(4, 1, 1, 1, 1, 1)$	1241
$(9, 10, 8, 4, 1, 1)$	2001	$(8, 1, 1, 1, 1, 1)$	1682
$(9, 9, 5, 8, 4, 1)$	1526	$(5, 1, 1, 1, 1, 1)$	2201
$(7, 4, 2, 2, 1, 1)$	2558	$(10, 1, 1, 1, 1, 1)$	3522
$(7, 8, 8, 8, 8, 1)$	1082	$(9, 1, 1, 1, 1, 1)$	1921
$(7, 7, 7, 4, 1, 1)$	3306	$(3, 1, 1, 1, 1, 1)$	2481
$(3, 8, 2, 2, 1, 1)$	2826	$(6, 1, 1, 1, 1, 1)$	1442
$(3, 5, 8, 4, 2, 1)$	1905	$(3, 10, 10, 8, 1, 1)$	3826
$(3, 7, 9, 5, 8, , 1)$	2001	$(3, 3, 3, 3, 8, 1)$	1058
$(6, 4, 4, 4, 4, 1)$	1130	$(6, 10, 8, 4, 4, 1)$	2626
$(6, 7, 10, 5, 4, 1)$	3018	$(6, 3, 3, 9, 10, 1)$	1925
$(6, 6, 6, 7, 7, 1)$	1294	$(4, 4, 1, 1, 1, 1)$	3261

c	$\rho_1(\mathcal{V}_k)$	c	$\rho_1(\mathcal{V}_k)$
$(8,4,1,1,1,1)$	1130	$(8,8,8,1,1,1)$	2294
$(5,5,2,1,1,1)$	1506	$(5,8,4,1,1,1)$	1478

A.3.3   Dimension 6. (a) We now compute the Picard numbers and the stable Picard numbers of diagonal hypersurfaces of dimension $n = 6$ with trivial twist and degree m, $4 \le m \le 25$, choosing $k = \mathbb{F}_q$ with $q = p \equiv 1 \pmod{m}$, so that all our motives are ordinary and the Tate conjecture is known to hold when m is prime and when $m \le 21$ (which covers all the cases we were able to compute).

m	p	$\rho_3(\mathcal{X}_k)$	$\bar{\rho}_3(\mathcal{V})$	m	p	$\rho_3(\mathcal{X}_k)$	$\bar{\rho}_3(\mathcal{V})$
4	5	492	1108	16	17		11328760
5	11	4901	4901	17	103	4649681	4649681
6	13	9158	38166	18	19		50590794
7	29	44731	44731	19	191	7792471	7792471
8	17	118400	219872	20	41		73206212
9	19	190121	343001	21	43		38846501
10	11	190794	1225450	22	23		60330622
11	23	551951	551951	23	47	18557771	18557771
12	13	765284	4882180	24	73		226084280
13	43	1281421	1281421	25	51		28377721
14	29	2719670	8729366	26	53		116261546
15	31	2570051	8970851				

For m prime, our computational results are consistent with the closed formula for the stable Picard number $\bar{\rho}_3(\mathcal{V})$ described in Proposition 5.4:

$$\bar{\rho}_3(\mathcal{V}) = 1 + 5 \cdot 7(m-1)(3m^3 - 27m^2 + 86m - 95),$$

which is obtained in much the same way as the one for dimension 4. See Appendix B.

(b) We compute the Picard numbers of diagonal hypersurfaces $\mathcal{V} = \mathcal{V}_6^m(\mathbf{c})$ for selected twists $\mathbf{c}$. Fix m, and choose a prime p such that $p \equiv 1 \pmod{m}$.

Pick a primitive root g modulo p. As above, we confine ourselves to twists $c = (c_0, c_1, \cdots, c_7)$ where each component runs over elements of the form g^j, $1 \leq j \leq m-1$. Again, extreme twists of the form $(g^j, 1, 1, 1, 1, 1, 1, 1)$, $1 \leq j \leq m-1$ will in general not be listed in the tables (since we know that the Picard number will be equal to 1 in those cases). We again observe that in all our examples twists of the form $(g^j, g^j, 1, 1, 1, 1, 1, 1)$ give the same Picard number for any j, $1 \leq j \leq m-1$ (and similarly, so do twists of the form $(g^j, g^j, g^j, 1, 1, 1, 1, 1)$).

(b1) Let $m = 5$, $n = 6$ and $p = 11$, and take $g = 2$. The stable Picard number is 4901.

c	$\rho_3(\mathcal{V}_k)$	c	$\rho_3(\mathcal{V}_k)$
$(2,2,1,1,1,1,1,1)$	1601	$(4,2,2,2,2,1,1,1)$	881
$(2,2,2,1,1,1,1,1)$	601	$(4,4,2,2,1,1,1,1)$	1081
$(4,2,1,1,1,1,1,1)$	1201	$(4,4,2,2,2,1,1,1)$	961
$(4,2,2,1,1,1,1,1)$	801	$(5,8,4,2,1,1,1,1)$	981
$(4,4,2,1,1,1,1,1)$	901	$(8,4,2,1,1,1,1,1)$	1001
$(2,2,2,2,1,1,1,1)$	1361	$(4,2,2,2,1,1,1,1)$	1061

(b2) Let $m = 7$, $n = 6$ and $p = 29$, and take $g = 2$. In this case the stable Picard number is 44731.

c	$\rho_3(\mathcal{V}_k)$	c	$\rho_3(\mathcal{V}_k)$
$(2,2,1,1,1,1,1,1)$	11161	$(16,8,4,4,2,1,1,1)$	6409
$(2,2,2,1,1,1,1,1)$	4141	$(16,8,4,4,4,1,1,1)$	6667
$(4,2,1,1,1,1,1,1)$	7201	$(16,8,8,4,2,1,1,1)$	6529
$(4,2,2,1,1,1,1,1)$	5341	$(16,16,8,2,1,1,1,1)$	6469
$(4,4,2,1,1,1,1,1)$	5701	$(16,16,8,2,2,1,1,1)$	6313
$(8,4,2,1,1,1,1,1)$	6121	$(16,16,8,8,1,1,1,1)$	7279
$(8,4,4,1,1,1,1,1)$	6151	$(16,16,8,8,4,1,1,1)$	6241
$(16,4,2,1,1,1,1,1)$	6931	$(16,16,16,4,2,1,1,1)$	6721
$(16,8,1,1,1,1,1,1)$	4771	$(3,8,4,2,1,1,1,1)$	6517
$(16,8,2,1,1,1,1,1)$	6541	$(3,8,4,2,2,1,1,1)$	6367
$(16,8,4,1,1,1,1,1)$	6901	$(3,8,4,4,2,1,1,1)$	6331

c	$\rho_3(\mathcal{V}_k)$	c	$\rho_3(\mathcal{V}_k)$
$(2,2,2,2,1,1,1,1)$	9061	$(3,16,4,4,2,1,1,1)$	6325
$(4,2,2,2,1,1,1,1)$	6757	$(3,16,8,8,4,1,1,1)$	6373
$(4,2,2,2,2,1,1,1)$	5569	$(3,3,4,4,4,1,1,1)$	6175
$(4,4,2,2,1,1,1,1)$	7357	$(3,3,8,8,1,1,1,1)$	7357
$(4,4,2,2,2,1,1,1)$	6175	$(6,16,8,4,2,1,1,1)$	6337
$(4,4,4,2,1,1,1,1)$	6343	$(6,3,8,8,2,1,1,1)$	6409
$(4,4,4,2,2,1,1,1)$	5851	$(6,6,6,8,8,1,1,1)$	5851
$(8,2,2,2,1,1,1,1)$	6613	$(6,6,6,2,1,1,1,1)$	5569
$(8,2,2,2,2,1,1,1)$	6253	$(8,4,2,2,1,1,1,1)$	6391
$(8,4,2,2,2,1,1,1)$	6379	$(8,4,4,2,1,1,1,1)$	6421
$(8,4,4,2,2,1,1,1)$	6403	$(8,8,2,2,1,1,1,1)$	7021
$(8,8,2,2,2,1,1,1)$	6193	$(8,8,4,2,1,1,1,1)$	6493
$(8,8,4,2,2,1,1,1)$	6301	$(8,8,4,4,2,1,1,1)$	6319
$(16,4,2,2,1,1,1,1)$	6325	$(16,4,2,2,2,1,1,1)$	6721
$(16,4,4,2,2,1,1,1)$	6217	$(16,4,4,4,1,1,1,1)$	6757
$(16,4,4,4,4,1,1,1)$	5569	$(16,8,2,2,1,1,1,1)$	6265
$(16,8,4,2,1,1,1,1)$	6277	$(16,8,4,2,2,1,1,1)$	6523
$(16,8,4,4,1,1,1,1)$	6097	$(16,16,1,1,1,1,1,1)$	11161

A.4 "Brauer numbers" of twisted Fermat motives

In this section, we shall compute the "Brauer numbers" of selected twisted Fermat motives for $\mathcal{V} = \mathcal{V}_n^m(\mathbf{c})$ with $n = 2d \geq 4$. When m is prime (and to some extent when m is an odd prime power), the norms of Jacobi sums (and hence the Brauer numbers) are well understood. However, when m is composite, these numbers are still mysterious. This is due, in part, to the fact that the Iwasawa type congruences are not known for composite cases. For composite cases, we list some of our computational results.

A.4.1 The case of m prime. Recall the definition of the "Brauer numbers" of twisted Fermat motives $\mathcal{V}_A$: if $\mathcal{V}_A$ is not supersingular, the norm of a twisted Jacobi sum is of the form

$$\text{Norm}_{L/\mathbb{Q}}(1 - \frac{\mathcal{J}(\mathbf{c}, \mathbf{a})}{q^d}) = \frac{B^d(\mathcal{V}_A) \cdot m}{q^{w_d(\mathcal{V}_A)}}$$

where $B_d(\mathcal{V}_A)$ is a square which may be divisible by m. If $\mathcal{V}_A$ is supersingular, but not strongly supersingular, the norm is equal to m. In the tables below, we list $B_d(\mathcal{V}_A)$ for non-supersingular $\mathcal{V}_A$, and $\frac{1}{m}\text{Norm}(1 - \frac{\mathcal{J}(\mathbf{c},\mathbf{a})}{q^d})$ for supersingular $\mathcal{V}_A$.

(a) Let $(m, n) = (13, 4)$. Choose $q = p = 53$. We take several characters (none supersingular), and tabulate their Brauer numbers $B_2(\mathcal{V}_A)$. The Milne–Lichtenbaum formula is known to hold in this case if the twist is extreme, as in the first column of the table; they are squares up to powers of m in general. Note that we get squares in all cases.

We use the following twisting vectors:

$$\mathbf{c}_1 = (17, 1, 1, 1, 1, 1) \qquad \mathbf{c}_2 = (15, 22, 11, 32, 4, 1)$$
$$\mathbf{c}_3 = (2, 2, 2, 2, 1, 1) \qquad \mathbf{c}_4 = (4, 2, 2, 2, 2, 1).$$

$\mathbf{a}$	$\mathbf{c}_1$	$\mathbf{c}_2$	$\mathbf{c}_3$	$\mathbf{c}_4$
$(1, 2, 3, 5, 6, 9)$	571^2	1	1663^2	79^2
$(1, 2, 4, 5, 7, 7)$	547^2	2^{12}	6007^2	1249^2
$(1, 1, 3, 4, 6, 11)$	5^4	1	1	53^2
$(1, 1, 2, 2, 3, 4)$	6473^2	2963^2	313^2	6473^2
$(1, 1, 1, 3, 9, 11)$	883^2	$79^2 \cdot 467^2$	$79^2 \cdot 467^2$	$53^2 \cdot 1117^2$
$(1, 1, 1, 5, 8, 10)$	25999^2	8969^2	233^2	571^2
$(1, 1, 1, 1, 3, 6)$	131^2	131^2	28807^2	233^2
$(1, 1, 1, 1, 1, 8)$	$181^2 \cdot 337^2$	$389^2 \cdot 1093^2$	$131^2 \cdot 8867^2$	$389^2 \cdot 1093^2$
$(1, 2, 5, 7, 12, 12)$	1	5^4	1	1
$(1, 4, 4, 6, 12, 12)$	131^2	547^2	181^2	181^2
$(1, 5, 10, 12, 12, 12)$	5^4	79^2	3^6	3^6

(b) Let $(m, n) = (7, 6)$. Choose $q = p = 29$. We take several characters (none supersingular) and twisting vectors

$$\mathbf{c_1} = (3,1,1,1,1,1,1,1) \qquad \mathbf{c_2} = (4,4,2,2,1,1,1,1)$$
$$\mathbf{c_3} = (8,4,4,2,2,1,1,1) \qquad \mathbf{c_4} = (6,6,3,16,1,1,1,1).$$

In each case, we tabulate the Brauer number $B_3(\mathcal{V}_A)$. Again, these numbers are known to be squares for extreme twists (first column); they are squares up to powers of m in general. Note that we get squares in all cases.

a	$\mathbf{c_1}$	$\mathbf{c_2}$	$\mathbf{c_3}$	$\mathbf{c_4}$
$(1,2,4,4,6,6,6,6)$	1	7^4	3^6	3^6
$(1,3,3,3,3,3,6,6)$	223^2	1	7^2	83^2
$(1,4,4,4,5,5,6,6)$	3^6	1	7^4	2^6
$(1,1,1,1,4,4,4,5)$	379^2	197^2	379^2	197^2
$(1,1,1,1,2,2,2,4)$	293^2	2143^2	113^2	$2^6 \cdot 83^2$

A.4.2 The case of m composite. For composite m, we do not have the Iwasawa congruence for twisted Jacobi sums, though when m is a power of an odd prime and is greater than 3, there seems to be some pattern. We know the denominator of each norm is of the form $q^{w(r)}$ for any r, $0 \le r \le n$. In the table below we list the numbers

$$q^{w(d)}\mathrm{Norm}_{L/\mathbb{Q}}\left(1 - \frac{\mathfrak{J}(\mathbf{c},\mathbf{a})}{q^d}\right)$$

for twisted Fermat motives $\mathcal{V}_A$ of even dimension $n = 2d$. As in Example 6.13 on page 71, observe that all primes of exact exponent 2 occurring in the tables are of the form ± 1 modulo some proper divisor d of m.

(a) Let $(m,n) = (9,4)$. Choose $q = p = 19$, and use the twisting vectors

$$\mathbf{c_1} = (17,1,1,1,1,1) \qquad \mathbf{c_2} = (15,3,11,13,4,1)$$
$$\mathbf{c_3} = (2,2,2,2,1,1) \qquad \mathbf{c_4} = (4,2,2,2,2,1).$$

a	c_1	c_2	c_3	c_4
$(1, 3, 8, 8, 8, 8)$	3	$3 \cdot 73^2$	$3 \cdot 19^2$	$3 \cdot 19^2$
$(1, 5, 6, 8, 8, 8)$	3	3^3	3	3
$(1, 2, 4, 4, 8, 8)$	$3 \cdot 37^2$	$3 \cdot 19^2$	$3 \cdot 17^2$	$3 \cdot 17^2$
$(1, 1, 1, 1, 1, 4)$	$3 \cdot 5^6$	$3 \cdot 71^2$	$3 \cdot 89^2$	3^3
$(1, 1, 1, 1, 7, 7)$	$3 \cdot 179^2$	$3 \cdot 179^2$	$3 \cdot 107^2$	3^5
$(1, 1, 2, 2, 6, 6)$	$3 \cdot 233^2$	$3 \cdot 17^2$	$3^3 \cdot 71^2$	$3 \cdot 2^6 \cdot 17^2$
$(1, 2, 6, 6, 6, 6)$	3	$3 \cdot 53^2$	3^3	$3 \cdot 37^2$
$(3, 3, 3, 3, 3, 3)$	$3^3 \cdot 5^6$	$3^3 \cdot 5^6$	$3^3 \cdot 5^6$	3^9

(b) Let $(m, n) = (20, 4)$. Choose $q = p = 41$, and twisting vectors

$$c_1 = (17, 1, 1, 1, 1, 1) \qquad c_2 = (15, 3, 11, 13, 4, 1)$$
$$c_3 = (2, 2, 2, 2, 1, 1) \qquad c_4 = (4, 3, 2, 2, 2, 1).$$

a	c_1	c_2	c_3	c_4
$(1, 3, 19, 19, 19, 19)$	79^2	2^4	41^2	61^2
$(1, 8, 14, 19, 19, 19)$	241^2	2^{12}	79^2	19^2
$(1, 4, 18, 19, 19, 19)$	461^2	59^2	1	59^2
$(1, 5, 17, 19, 19, 19)$	419^2	5^4	$2^4 \cdot 5^2$	$2^4 \cdot 101^2$
$(1, 11, 11, 19, 19, 19)$	359^2	379^2	379^2	359^2
$(1, 3, 3, 15, 19, 19)$	1	2^4	41^2	2^4
$(1, 2, 8, 11, 19, 19)$	461^2	2^8	61^2	2^{12}
$(1, 4, 5, 12, 19, 19)$	1	$2^4 \cdot 5^2$	$2^4 \cdot 5^2$	5^4
$(1, 3, 9, 9, 19, 19)$	79^2	$2^4 \cdot 5^2 \cdot 19^2$	$5^2 \cdot 61^2$	61^2
$(4, 4, 4, 4, 12, 12)$	$5^2 \cdot 31^4$	$5^2 \cdot 59^4$	5^2	$5^2 \cdot 59^4$
$(4, 4, 4, 5, 8, 15)$	5^2	2^4	11^4	3^8
$(4, 4, 5, 5, 10, 12)$	$2^8 \cdot 19^2$	61^2	41^2	41^2
$(4, 5, 5, 5, 5, 16)$	619^2	1439^2	$5^2 \cdot 419^2$	1439^2

A.5 Global "Brauer numbers" of $\mathcal{V} = \mathcal{V}_n^m(\mathbf{c})$

Finally we are ready to compute the global "Brauer numbers" of diagonal hypersurfaces $\mathcal{V} = \mathcal{V}_n^m(\mathbf{c})$.

A.5.1 The case of prime degree. (a) Let $(m, n) = (5, 4)$. We choose $q = p = 11$ (the smallest prime congruent to 1 modulo 5, for simplicity), and consider twists $\mathbf{c} = (c_0, c_1, c_2, c_3, c_4, c_5)$. We compute the "Brauer numbers" of all twisted Fermat motives for $\mathcal{V} = \mathcal{V}_4^5(\mathbf{c})$. There are altogether 820 characters $\mathbf{a} \in \mathfrak{A}_4^5$. For the trivial twist $\mathbf{c} = \mathbf{1}$, they are divided into 5 non-isomorphic types of motives, all of dimension 4. The isomorphism types are represented by the characters $[1, 3, 4, 4, 4, 4]$, $[1, 1, 1, 4, 4, 4]$, $[1, 1, 2, 3, 4, 4]$, $[1, 2, 2, 2, 4, 4]$, and $[1, 1, 1, 1, 3, 3]$. We know that $\rho_2(\mathcal{X}_k) = 401$.

If we introduce a non-trivial twist $\mathbf{c}$, each equivalence class is further divided into subclasses. We compute the "Brauer numbers" for Fermat motives with twists $\mathbf{c}_1 = (2, 1, 1, 1, 1, 1)$. In this case the non-isomorphic subclasses are distinguished simply by their a_0 components, which we list in the table.

Since we have $p \equiv 1 \pmod 5$, all motives are ordinary; the second and third motives are ordinary and supersingular. These are distinguished by an asterisk in the tables. For completeness, we have also included the strongly supersingular motives, which can be recognized by the fact that they have 0 in the column for the "Brauer number" (because the norm is zero in that case).

The table below lists the "Brauer numbers" $B^2(\mathcal{M}_A)$ and $B^2(\mathcal{V}_A)$.

	$\mathbf{a}$	$w(2)$	a_0	mult.	$B^2(\mathcal{M}_A)$	$B^2(\mathcal{V}_A)$
1a	$(1, 3, 4, 4, 4, 4)$	1	1	20	5^2	2^4
1b	$(1, 3, 4, 4, 4, 4)$	1	3	20	5^2	1
1c	$(1, 3, 4, 4, 4, 4)$	1	4	80	5^2	1
2a	$*(1, 1, 1, 4, 4, 4)$	0	1	20	0	1
2b	$*(1, 1, 1, 4, 4, 4)$	0	4	20	0	1
3a	$*(1, 1, 2, 3, 4, 4)$	0	1	120	0	1
3b	$*(1, 1, 2, 3, 4, 4)$	0	2	60	0	1
3c	$*(1, 1, 2, 3, 4, 4)$	0	3	60	0	1
3d	$*(1, 1, 2, 3, 4, 4)$	0	4	120	0	1
4a	$(1, 2, 2, 2, 4, 4)$	1	1	40	5^2	1
4b	$(1, 2, 2, 2, 4, 4)$	1	2	120	5^2	1
4c	$(1, 2, 2, 2, 4, 4)$	1	4	80	5^2	1

	a	$w(2)$	a_0	mult.	$B^2(\mathcal{M}_A)$	$B^2(\mathcal{V}_A)$
5a	$(1,1,1,1,3,3)$	2	1	40	5^2	2^4
5b	$(1,1,1,1,3,3)$	2	3	20	5^2	19^2

In this case the global "Brauer numbers" are:

$$B^2(\mathcal{X}_k) = 5^{840} \quad \text{and} \quad B^2(\mathcal{V}_k) = 2^{240} \cdot 19^{40}.$$

(b) Now choose $q = p = 11$ and let the twist be $\mathbf{c} = (2,2,1,1,1,1)$. The corresponding Picard number $\rho_2(\mathcal{V}_k)$ is 145. The isomorphism classes are now divided into subclasses corresponding to the first two entries a_0, a_1. We list the "Brauer numbers" $B^2(\mathcal{V}_A)$.

	a	$w(2)$	a_0, a_1	mult.	$B^2(\mathcal{V}_A)$
1a	$(1,3,4,4,4,4)$	1	1,3	4	1
1b	$(1,3,4,4,4,4)$	1	1,4	16	5^2
1c	$(1,3,4,4,4,4)$	1	3,1	4	1
1d	$(1,3,4,4,4,4)$	1	3,4	16	1
1e	$(1,3,4,4,4,4)$	1	4,1	16	5^2
1f	$(1,3,4,4,4,4)$	1	4,3	16	1
1g	$(1,3,4,4,4,4)$	1	4,4	48	1
2a	$*(1,1,1,4,4,4)$	0	1,1	8	1
2b	$*(1,1,1,4,4,4)$	0	1,4	12	0
2c	$*(1,1,1,4,4,4)$	0	4,1	12	0
2d	$*(1,1,1,4,4,4)$	0	4,4	8	1
3a	$*(1,1,2,3,4,4)$	0	1,1	24	1
3b	$*(1,1,2,3,4,4)$	0	1,2	24	1
3c	$*(1,1,2,3,4,4)$	0	1,3	24	1
3d	$*(1,1,2,3,4,4)$	0	1,4	48	0
3e	$*(1,1,2,3,4,4)$	0	2,1	24	1
3f	$*(1,1,2,3,4,4)$	0	2,3	12	0
3g	$*(1,1,2,3,4,4)$	0	2,4	24	1
3h	$*(1,1,2,3,4,4)$	0	3,1	24	1

	a	$w(2)$	a_0, a_1	mult.	$B^2(\mathcal{V}_A)$
$3i$	$*(1,1,2,3,4,4)$	0	3,2	12	0
$3j$	$*(1,1,2,3,4,4)$	0	3,4	24	1
$3k$	$*(1,1,2,3,4,4)$	0	4,1	48	0
$3l$	$*(1,1,2,3,4,4)$	0	4,2	24	1
$3m$	$*(1,1,2,3,4,4)$	0	4,3	24	1
$3n$	$*(1,1,2,3,4,4)$	0	4,4	24	1
$4a$	$(1,2,2,2,4,4)$	1	1,2	24	2^4
$4b$	$(1,2,2,2,4,4)$	1	1,4	16	5^2
$4c$	$(1,2,2,2,4,4)$	1	2,1	24	2^4
$4d$	$(1,2,2,2,4,4)$	1	2,2	48	1
$4e$	$(1,2,2,2,4,4)$	1	2,4	48	1
$4f$	$(1,2,2,2,4,4)$	1	4,1	16	5^2
$4g$	$(1,2,2,2,4,4)$	1	4,2	48	1
$4h$	$(1,2,2,2,4,4)$	1	4,4	16	2^4
$5a$	$(1,1,1,1,3,3)$	2	1,1	24	1
$5b$	$(1,1,1,1,3,3)$	2	1,3	16	3^4
$5c$	$(1,1,1,1,3,3)$	2	3,1	16	3^4
$5d$	$(1,1,1,1,3,3)$	2	3,3	4	2^4

The global "Brauer number" is $B^2(\mathcal{V}_k) = 2^{272} \cdot 3^{128} \cdot 5^{128}$.

(c) Let $(m, n) = (7, 4)$ Here we choose $q = p = 29$ and twists $\mathbf{c} = \mathbf{1}$ and $\mathbf{c} = (3, 1, 1, 1, 1, 1)$. We compute the "Brauer numbers" of all twisted Fermat motives for $\mathcal{V} = \mathcal{V}_4^7(\mathbf{c})$, and the global "Brauer number" $B^2(\mathcal{V}_k)$. There are altogether $6,666$ characters $\mathbf{a} \in \mathfrak{A}_4^7$. For the trivial twist $\mathbf{c} = \mathbf{1}$, one needs to consider 14 representatives of isomorphism classes of Fermat motives. If we have the extreme twist $\mathbf{c}$ as above, each isomorphism class is further divided into subclasses which are distinguished by their a_0 entry. We compute the "Brauer numbers" for the Fermat motives and for the twisted Fermat motives with twist $\mathbf{c} = (3, 1, 1, 1, 1, 1)$. Again with our choice of p with $p \equiv 1 \pmod{7}$, all motives are ordinary; motives which are ordinary and supersingular are indicated by an asterisk.

	a	$w(2)$	a_0	mult.	$B^2(\mathcal{M}_A)$	$B^2(\mathcal{V}_A)$
1a	$(1,3,6,6,6,6)$	2	1	30	7^2	41^2
1b	$(1,3,6,6,6,6)$	2	3	30	7^2	41^2
1c	$(1,3,6,6,6,6)$	2	6	120	7^2	2^6
2a	$*(1,1,1,6,6,6)$	0	1	30	0	1
2b	$*(1,1,1,6,6,6)$	0	6	30	0	1
3a	$(1,4,5,6,6,6)$	1	1	120	7^2	1
3b	$(1,4,5,6,6,6)$	1	4	120	7^2	2^6
3c	$(1,4,5,6,6,6)$	1	5	120	7^2	13^2
3d	$(1,4,5,6,6,6)$	1	6	360	7^2	1
4a	$*(1,1,2,5,6,6)$	0	1	180	0	1
4b	$*(1,1,2,5,6,6)$	0	2	90	0	1
4c	$*(1,1,2,5,6,6)$	0	5	90	0	1
4d	$*(1,1,2,5,6,6)$	0	6	180	0	1
5a	$*(1,1,3,4,6,6)$	0	1	180	0	1
5b	$*(1,1,3,4,6,6)$	0	3	90	0	1
5c	$*(1,1,3,4,6,6)$	0	4	90	0	1
5d	$*(1,1,3,4,6,6)$	0	6	180	0	1
6a	$(1,2,2,4,6,6)$	1	1	180	7^2	1
6b	$(1,2,2,4,6,6)$	1	2	360	7^2	1
6c	$(1,2,2,4,6,6)$	1	4	180	7^2	13^2
6d	$(1,2,2,4,6,6)$	1	6	360	7^2	2^6
7a	$(1,2,3,3,6,6)$	1	1	180	7^2	1
7b	$(1,2,3,3,6,6)$	1	2	180	7^2	2^6
7c	$(1,2,3,3,6,6)$	1	3	360	7^2	1
7d	$(1,2,3,3,6,6)$	1	6	360	7^2	13^2
8a	$(1,5,5,5,6,6)$	2	1	60	7^2	1
8b	$(1,5,5,5,6,6)$	2	5	180	7^2	2^6
8c	$(1,5,5,5,6,6)$	2	6	120	7^2	41^2
9a	$(1,1,1,1,1,2)$	3	1	30	7^2	83^2
9b	$(1,1,1,1,1,2)$	3	2	6	7^2	43^2

	a	$w(2)$	a_0	mult.	$B^2(\mathcal{M}_A)$	$B^2(\mathcal{V}_A)$
10a	$(1,1,1,1,5,5)$	3	1	60	7^2	83^2
10b	$(1,1,1,1,5,5)$	3	5	30	7^2	13^2
11a	$(1,1,1,2,4,5)$	2	1	360	7^2	2^6
11b	$(1,1,1,2,4,5)$	2	2	120	7^2	13^2
11c	$(1,1,1,2,4,5)$	2	4	120	7^2	41^2
11d	$(1,1,1,2,4,5)$	2	5	120	7^2	43^2
12a	$(1,1,1,3,3,5)$	2	1	180	7^4	41^2
12b	$(1,1,1,3,3,5)$	2	3	120	7^4	1
12c	$(1,1,1,3,3,5)$	2	5	60	7^4	13^2
13a	$(1,1,2,2,4,4)$	3	1	60	$2^{12} \cdot 7^2$	71^2
13b	$(1,1,2,2,4,4)$	3	2	60	$2^{12} \cdot 7^2$	71^2
13c	$(1,1,2,2,4,4)$	3	4	60	$2^{12} \cdot 7^2$	71^2
14a	$*(1,2,3,4,5,6)$	0	1	120	0	1
14b	$*(1,2,3,4,5,6)$	0	2	120	0	1
14c	$*(1,2,3,4,5,6)$	0	3	120	0	1
14d	$*(1,2,3,4,5,6)$	0	4	120	0	1
14e	$*(1,2,3,4,5,6)$	0	5	120	0	1
14f	$*(1,2,3,4,5,6)$	0	6	120	0	1

We obtain the global "Brauer numbers"

$$B^2(\mathcal{X}_k) = 2^{2160} \cdot 7^{10332}$$

and

$$B^2(\mathcal{V}_k) = 2^{7920} \cdot 13^{1740} \cdot 41^{960} \cdot 43^{252} \cdot 71^{360} \cdot 83^{180}.$$

(d) Let $(m,n) = (7,6)$. We choose $p = 29$, and compute the global "Brauer number" of $\mathcal{V} = \mathcal{V}_6^7(\mathbf{c})$ over $\mathbb{F}_{29}$ for $\mathbf{c} = \mathbf{1}$ and for $\mathbf{c} = (2,1,1,1,1,1,1,1)$. We compute "Brauer numbers" of all the twisted Fermat motives in each case. In the tables, $w(3)$ and $B^3(\mathcal{V}_A)$ are as defined in Theorem 6.2. Supersingular motives are indicated by an asterisk.

	a	$w(3)$	a_0	mult.	$B^3(\mathcal{M}_A)$	$B^3(\mathcal{V}_A)$
1a	$(1,5,6,6,6,6,6,6)$	3	1	42	7^2	2^6
1b	$(1,5,6,6,6,6,6,6)$	3	5	42	7^2	5^6
1c	$(1,5,6,6,6,6,6,6)$	3	6	252	7^2	$13^2 \cdot 29^2$
2a	$(1,1,3,6,6,6,6,6)$	2	1	252	7^2	41^2
2b	$(1,1,3,6,6,6,6,6)$	2	3	126	7^2	43^2
2c	$(1,1,3,6,6,6,6,6)$	2	6	630	7^2	1
3a	$(1,2,2,6,6,6,6,6)$	3	1	126	7^2	349^2
3b	$(1,2,2,6,6,6,6,6)$	3	2	252	7^2	83^2
3c	$(1,2,2,6,6,6,6,6)$	3	6	630	7^2	281^2
4a	$*(1,1,1,1,6,6,6,6)$	0	1	105	0	1
4b	$*(1,1,1,1,6,6,6,6)$	0	6	105	0	1
5a	$(1,1,4,5,6,6,6,6)$	1	1	1260	7^2	1
5b	$(1,1,4,5,6,6,6,6)$	1	4	630	7^2	13^2
5c	$(1,1,4,5,6,6,6,6)$	1	5	630	7^2	1
5d	$(1,1,4,5,6,6,6,6)$	1	6	2520	7^2	2^6
6a	$(1,2,3,5,6,6,6,6)$	2	1	1260	7^2	41^2
6b	$(1,2,3,5,6,6,6,6)$	2	2	1260	7^2	2^6
6c	$(1,2,3,5,6,6,6,6)$	2	3	1260	7^2	43^2
6d	$(1,2,3,5,6,6,6,6)$	2	5	1260	7^2	41^2
6e	$(1,2,3,5,6,6,6,6)$	2	6	5040	7^2	1
7a	$(1,2,4,4,6,6,6,6)$	2	1	630	7^4	2^6
7b	$(1,2,4,4,6,6,6,6)$	2	2	630	7^4	41^2
7c	$(1,2,4,4,6,6,6,6)$	2	4	1260	7^4	3^6
7d	$(1,2,4,4,6,6,6,6)$	2	6	2520	7^4	1
8a	$(1,3,3,4,6,6,6,6)$	2	1	630	7^2	41^2
8b	$(1,3,3,4,6,6,6,6)$	2	3	1260	7^2	43^2
8c	$(1,3,3,4,6,6,6,6)$	2	4	630	7^2	13^2
8d	$(1,3,3,4,6,6,6,6)$	2	6	2520	7^2	1
9a	$*(1,1,1,2,5,6,6,6)$	0	1	1260	0	1
9b	$*(1,1,1,2,5,6,6,6)$	0	2	420	0	1

	a	$w(3)$	a_0	mult.	$B^3(\mathcal{M}_A)$	$B^3(\mathcal{V}_A)$
9c	$*(1,1,1,2,5,6,6,6)$	0	5	420	0	1
9d	$*(1,1,1,2,5,6,6,6)$	0	6	1260	0	1
10a	$*(1,1,1,3,4,6,6,6)$	0	1	1260	0	1
10b	$*(1,1,1,3,4,6,6,6)$	0	3	420	0	1
10c	$*(1,1,1,3,4,6,6,6)$	0	4	420	0	1
10d	$*(1,1,1,3,4,6,6,6)$	0	6	1260	0	1
11a	$(1,1,2,2,4,6,6,6)$	1	1	2520	7^2	1
11b	$(1,1,2,2,4,6,6,6)$	1	2	2520	7^2	2^6
11c	$(1,1,2,2,4,6,6,6)$	1	4	1260	7^2	1
11d	$(1,1,2,2,4,6,6,6)$	1	6	3780	7^2	13^2
12a	$(1,1,2,3,3,6,6,6)$	1	1	2520	7^2	1
12b	$(1,1,2,3,3,6,6,6)$	1	2	1260	7^2	13^2
12c	$(1,1,2,3,3,6,6,6)$	1	3	2520	7^2	2^6
12d	$(1,1,2,3,3,6,6,6)$	1	6	3780	7^2	1
13a	$(1,1,5,5,5,6,6,6)$	2	1	840	7^2	13^2
13b	$(1,1,5,5,5,6,6,6)$	2	5	1260	7^2	1
13c	$(1,1,5,5,5,6,6,6)$	2	6	1260	7^2	43^2
14a	$(1,2,2,2,3,6,6,6)$	2	1	840	7^4	13^2
14b	$(1,2,2,2,3,6,6,6)$	2	2	2520	7^4	1
14c	$(1,2,2,2,3,6,6,6)$	2	3	840	7^4	41^2
14d	$(1,2,2,2,3,6,6,6)$	2	6	2520	7^4	3^6
15a	$(1,2,4,5,5,6,6,6)$	1	1	2520	7^2	1
15b	$(1,2,4,5,5,6,6,6)$	1	2	2520	7^2	1
15c	$(1,2,4,5,5,6,6,6)$	1	4	2520	7^2	13^2
15d	$(1,2,4,5,5,6,6,6)$	1	5	5040	7^2	1
15e	$(1,2,4,5,5,6,6,6)$	1	6	7560	7^2	2^6
16a	$(1,3,3,5,5,6,6,6)$	3	1	1260	$2^{12} \cdot 7^2$	71^2
16b	$(1,3,3,5,5,6,6,6)$	3	3	2520	$2^{12} \cdot 7^2$	13^2
16c	$(1,3,3,5,5,6,6,6)$	3	5	2520	$2^{12} \cdot 7^2$	13^2
16d	$(1,3,3,5,5,6,6,6)$	3	6	3780	$2^{12} \cdot 7^2$	13^2

	a	$w(3)$	a_0	mult.	$B^3(\mathcal{M}_A)$	$B^3(\mathcal{V}_A)$
17a	$(1,3,4,4,5,6,6,6)$	1	1	2520	7^2	1
17b	$(1,3,4,4,5,6,6,6)$	1	3	2520	7^2	1
17c	$(1,3,4,4,5,6,6,6)$	1	4	5040	7^2	13^2
17d	$(1,3,4,4,5,6,6,6)$	1	5	2520	7^2	1
17e	$(1,3,4,4,5,6,6,6)$	1	6	7560	7^2	2^6
18a	$(1,4,4,4,4,6,6,6)$	3	1	210	7^2	13^4
18b	$(1,4,4,4,4,6,6,6)$	3	4	840	7^2	281^2
18c	$(1,4,4,4,4,6,6,6)$	3	6	630	7^2	83^2
19a	$*(1,1,2,2,5,5,6,6)$	0	1	1890	0	1
19b	$*(1,1,2,2,5,5,6,6)$	0	2	1890	0	1
19c	$*(1,1,2,2,5,5,6,6)$	0	5	1890	0	1
19d	$*(1,1,2,2,5,5,6,6)$	0	6	1890	0	1
20a	$*(1,1,2,3,4,5,6,6)$	0	1	7560	0	1
20b	$*(1,1,2,3,4,5,6,6)$	0	2	3780	0	1
20c	$*(1,1,2,3,4,5,6,6)$	0	3	3780	0	1
20d	$*(1,1,2,3,4,5,6,6)$	0	4	3780	0	1
20e	$*(1,1,2,3,4,5,6,6)$	0	5	3780	0	1
20f	$*(1,1,2,3,4,5,6,6)$	0	6	7560	0	1
21a	$(1,1,2,4,4,4,6,6)$	2	1	2520	7^2	41^2
21b	$(1,1,2,4,4,4,6,6)$	2	2	1260	7^2	43^2
21c	$(1,1,2,4,4,4,6,6)$	2	4	3780	7^2	1
21d	$(1,1,2,4,4,4,6,6)$	2	6	2520	7^2	2^6
22a	$(1,2,2,3,3,5,6,6)$	1	1	3780	7^2	1
22b	$(1,2,2,3,3,5,6,6)$	1	2	7560	7^2	13^2
22c	$(1,2,2,3,3,5,6,6)$	1	3	7560	7^2	2^6
22d	$(1,2,2,3,3,5,6,6)$	1	5	3780	7^2	1
22e	$(1,2,2,3,3,5,6,6)$	1	6	7560	7^2	1
23a	$(1,3,3,3,3,3,6,6)$	3	1	126	7^2	83^2
23b	$(1,3,3,3,3,3,6,6)$	3	3	630	7^2	$13^2 \cdot 29^2$
23c	$(1,3,3,3,3,3,6,6)$	3	6	252	7^2	5^6

	a	$w(3)$	a_0	mult.	$B^3(\mathcal{M}_A)$	$B^3(\mathcal{V}_A)$
24a	$(1,3,4,5,5,5,6,6)$	2	1	2520	7^2	13^2
24b	$(1,3,4,5,5,5,6,6)$	2	3	2520	7^2	41^2
24c	$(1,3,4,5,5,5,6,6)$	2	4	2520	7^2	2^6
24d	$(1,3,4,5,5,5,6,6)$	2	5	7560	7^2	1
24e	$(1,3,4,5,5,5,6,6)$	2	6	5040	7^2	43^2
25a	$(1,4,4,4,5,5,6,6)$	2	1	1260	7^4	1
25b	$(1,4,4,4,5,5,6,6)$	2	4	3780	7^4	1
25c	$(1,4,4,4,5,5,6,6)$	2	5	2520	7^4	3^6
25d	$(1,4,4,4,5,5,6,6)$	2	6	2520	7^4	41^2
26a	$(1,1,1,1,1,1,3,5)$	4	1	252	$7^2 \cdot 197^2$	953^2
26b	$(1,1,1,1,1,1,3,5)$	4	3	42	$7^2 \cdot 197^2$	631^2
26c	$(1,1,1,1,1,1,3,5)$	4	5	42	$7^2 \cdot 197^2$	$13^2 \cdot 71^2$
27a	$(1,1,1,1,1,1,4,4)$	4	1	126	$7^2 \cdot 97^2$	43^2
27b	$(1,1,1,1,1,1,4,4)$	4	4	42	$7^2 \cdot 97^2$	$41^2 \cdot 43^2$
28a	$(1,1,1,1,1,2,3,4)$	3	1	1260	7^2	$13^2 \cdot 29^2$
28b	$(1,1,1,1,1,2,3,4)$	3	2	252	7^2	5^6
28c	$(1,1,1,1,1,2,3,4)$	3	3	252	7^2	43^2
28d	$(1,1,1,1,1,2,3,4)$	3	4	252	7^2	223^2
29a	$(1,1,1,1,1,3,3,3)$	3	1	210	$2^6 \cdot 7^2$	43^2
29b	$(1,1,1,1,1,3,3,3)$	3	3	126	$2^6 \cdot 7^2$	1
30a	$(1,1,1,1,2,2,2,4)$	4	1	840	$7^2 \cdot 41^2$	$2^6 \cdot 83^2$
30b	$(1,1,1,1,2,2,2,4)$	4	2	630	$7^2 \cdot 41^2$	461^2
30c	$(1,1,1,1,2,2,2,4)$	4	4	210	$7^2 \cdot 41^2$	$3^6 \cdot 13^2$
31a	$(1,1,1,1,2,2,3,3)$	2	1	1260	7^2	13^2
31b	$(1,1,1,1,2,2,3,3)$	2	2	630	7^2	2^{12}
31c	$(1,1,1,1,2,2,3,3)$	2	3	630	7^2	13^2
32a	$(1,1,1,1,3,4,5,5)$	3	1	2520	7^2	281^2
32b	$(1,1,1,1,3,4,5,5)$	3	3	630	7^2	2^{12}
32c	$(1,1,1,1,3,4,5,5)$	3	4	630	7^2	13^2
32d	$(1,1,1,1,3,4,5,5)$	3	5	1260	7^2	83^2

	a	$w(3)$	a_0	mult.	$B^3(\mathcal{M}_A)$	$B^3(\mathcal{V}_A)$
33a	$(1,1,1,1,4,4,4,5)$	3	1	840	$7^2 \cdot 13^2$	97^2
33b	$(1,1,1,1,4,4,4,5)$	3	4	630	$7^2 \cdot 13^2$	1
33c	$(1,1,1,1,4,4,4,5)$	3	5	210	$7^2 \cdot 13^2$	379^2

Observe that the prime factors of $B^3(\mathcal{V}_A)$ with the exception of 2, 3 and 5 are all of the form $7k \pm 1$.

Putting together all these motivic "Brauer numbers" counted with correct multiplicities, we obtain, as pointed out in the main text,

$$
\begin{aligned}
B^3(\mathcal{V}_k) =\; & 2^{152220} \cdot 3^{25200} \cdot 5^{2268} \cdot 13^{53056} \cdot 29^{3024} \cdot 41^{16576} \cdot 43^{14392} \cdot \\
& 71^{1736} \cdot 83^{4144} \cdot 97^{1120} \cdot 223^{336} \cdot 281^{5320} \cdot 349^{168} \cdot 379^{280} \cdot \\
& 461^{840} \cdot 631^{56} \cdot 953^{336}.
\end{aligned}
$$

A.5.2   The case of composite degree. Finally, we do a few computations in the composite case. Since we do not, for m composite, have precise information as to the m-part of the norm, we prefer to record in each case the value of

$$
p^{w(d)} \operatorname{Norm}\left(1 - \frac{\mathfrak{J}(\mathbf{c}, \mathbf{a})}{p^d}\right).
$$

There are two cases to consider. First, if $m = m_0^r$ is an odd prime power, the Iwasawa congruence is known to hold, and therefore contributes to the norm an odd power of the prime m_0. We can observe this below, in the case $m = 9$.

If, on the other hand, m is not an odd prime power, things seem to be much less clear. In particular, we list below the results for $m = 6$, $n = 4$. In this case we see that the norm is sometimes divisible by 2, sometimes by 3, sometimes by both, sometimes by neither. The prime 2 always occurs to an even power, while 3, when it occurs, appears with an odd power.

(a) Let $m = 9$, $n = 2$, $p = 19$. In this case the twist $\mathbf{c} = (4,1,1,1)$ is extreme, in the sense that $\rho(\mathcal{V}_k) = 1$. The following table records the values of

$$
p^{w(1)} \operatorname{Norm}\left(1 - \frac{\mathfrak{J}(\mathbf{c}, \mathbf{a})}{p}\right)
$$

for the trivial twist $\mathbf{c}_1 = 1$ and for $\mathbf{c}_2 = (4,1,1,1)$, with $\mathbf{a}$ running through a list of representatives of the isomorphism classes of twisted Fermat motives. As usual, we break the nine isomorphism classes of Fermat motives into subclasses determined by the first coefficient, a_0. Supersingular motives are marked by an asterisk.

	a	$w(1)$	a_0	mult.	$\mathbf{c_1 = 1}$	$\mathbf{c_2}$
$1a$	$*(1,1,8,8)$	0	1	9	0	3
$1b$	$*(1,1,8,8)$	0	8	9	0	3
$2a$	$(1,1,1,6)$	2	1	18	$2^6 \cdot 3^3$	$3 \cdot 17^2$
$2b$	$(1,1,1,6)$	2	6	6	$2^6 \cdot 3^3$	3^3
$3a$	$(1,1,2,5)$	2	1	36	3^7	3
$3b$	$(1,1,2,5)$	2	2	18	3^7	3
$3c$	$(1,1,2,5)$	2	5	18	3^7	$3 \cdot 19^2$
$4a$	$(1,1,3,4)$	1	1	36	3^3	$2^6 \cdot 3$
$4b$	$(1,1,3,4)$	1	3	18	3^3	3^5
$4c$	$(1,1,3,4)$	1	4	18	3^3	3
$5a$	$(1,2,3,3)$	1	1	18	3^5	3
$5b$	$(1,2,3,3)$	1	2	18	3^5	$3 \cdot 17^2$
$5c$	$(1,2,3,3)$	1	3	36	3^5	3^3
$6a$	$*(1,2,7,8)$	0	1	18	0	3
$6b$	$*(1,2,7,8)$	0	2	18	0	3
$6c$	$*(1,2,7,8)$	0	7	18	0	3
$6d$	$*(1,2,7,8)$	0	8	18	0	3
$7a$	$*(1,3,6,8)$	0	1	18	0	3
$7b$	$*(1,3,6,8)$	0	3	18	0	3^3
$7c$	$*(1,3,6,8)$	0	6	18	0	3^3
$7d$	$*(1,3,6,8)$	0	8	18	0	3
$8a$	$*(1,4,6,7)$	0	1	12	3^3	3
$8b$	$*(1,4,6,7)$	0	4	12	3^3	3
$8c$	$*(1,4,6,7)$	0	6	12	3^3	3^3
$8d$	$*(1,4,6,7)$	0	7	12	3^3	3
$9a$	$*(3,3,6,6)$	0	3	3	0	3^3
$9b$	$*(3,3,6,6)$	0	6	3	0	3^3

(b) Let $m = 9$, $n = 6$, $p = 19$. The following table records the values of

$$p^{w(3)} \operatorname{Norm}\left(1 - \frac{\vartheta(\mathbf{c}, \mathbf{a})}{p^3}\right)$$

for $c_1 = 1$ and $c_2 = (4, 1, 1, 1, 1, 1, 1, 1)$ and **a** running through a list of representatives of the isomorphism classes of twisted Fermat motives. As usual, we break the 129 isomorphism classes of Fermat motives into subclasses determined by the first coefficient, a_0. Supersingular motives are marked by an asterisk.

	a	$w(3)$	a_0	mult.	$c_1 = 1$	c_2
1a	$(1, 5, 8, 8, 8, 8, 8, 8)$	3	1	42	$3^3 \cdot 109^2$	$3 \cdot 71^2$
1b	$(1, 5, 8, 8, 8, 8, 8, 8)$	3	5	42	$3^3 \cdot 109^2$	$3 \cdot 53^2$
1c	$(1, 5, 8, 8, 8, 8, 8, 8)$	3	8	252	$3^3 \cdot 109^2$	$2^6 \cdot 3$
2a	$(1, 1, 3, 8, 8, 8, 8, 8)$	2	1	252	$2^6 \cdot 3^3$	$3 \cdot 19^2$
2b	$(1, 1, 3, 8, 8, 8, 8, 8)$	2	3	126	$2^6 \cdot 3^3$	3^3
2c	$(1, 1, 3, 8, 8, 8, 8, 8)$	2	8	630	$2^6 \cdot 3^3$	$3 \cdot 17^2$
3a	$(1, 2, 2, 8, 8, 8, 8, 8)$	3	1	126	$3^3 \cdot 17^2$	3
3b	$(1, 2, 2, 8, 8, 8, 8, 8)$	3	2	252	$3^3 \cdot 17^2$	$2^6 \cdot 3 \cdot 17^2$
3c	$(1, 2, 2, 8, 8, 8, 8, 8)$	3	8	630	$3^3 \cdot 17^2$	$3 \cdot 17^2$
4a	$(1, 6, 7, 8, 8, 8, 8, 8)$	3	1	252	3^3	$3 \cdot 89^2$
4b	$(1, 6, 7, 8, 8, 8, 8, 8)$	3	6	252	3^3	$3^5 \cdot 17^2$
4c	$(1, 6, 7, 8, 8, 8, 8, 8)$	3	7	252	3^3	$3 \cdot 5^6$
4d	$(1, 6, 7, 8, 8, 8, 8, 8)$	3	8	1260	3^3	$3 \cdot 53^2$
5a	$*(1, 1, 1, 1, 8, 8, 8, 8)$	0	1	105	0	3
5b	$*(1, 1, 1, 1, 8, 8, 8, 8)$	0	8	105	0	3
6a	$(1, 1, 4, 7, 8, 8, 8, 8)$	2	1	1260	3^7	$3 \cdot 37^2$
6b	$(1, 1, 4, 7, 8, 8, 8, 8)$	2	4	630	3^7	$3 \cdot 19^2$
6c	$(1, 1, 4, 7, 8, 8, 8, 8)$	2	7	630	3^7	3
6d	$(1, 1, 4, 7, 8, 8, 8, 8)$	2	8	2520	3^7	3
7a	$(1, 1, 5, 6, 8, 8, 8, 8)$	1	1	1260	3^3	3
7b	$(1, 1, 5, 6, 8, 8, 8, 8)$	1	5	630	3^3	3
7c	$(1, 1, 5, 6, 8, 8, 8, 8)$	1	6	630	3^3	3^5
7d	$(1, 1, 5, 6, 8, 8, 8, 8)$	1	8	2520	3^3	$2^6 \cdot 3$
8a	$(1, 2, 3, 7, 8, 8, 8, 8)$	2	1	1260	$2^6 \cdot 3^3$	$3 \cdot 19^2$
8b	$(1, 2, 3, 7, 8, 8, 8, 8)$	2	2	1260	$2^6 \cdot 3^3$	$3 \cdot 73^2$
8c	$(1, 2, 3, 7, 8, 8, 8, 8)$	2	3	1260	$2^6 \cdot 3^3$	3^3

	a	$w(3)$	a_0	mult.	$\mathbf{c_1 = 1}$	$\mathbf{c_2}$
$8d$	$(1,2,3,7,8,8,8,8)$	2	7	1260	$2^6 \cdot 3^3$	$3 \cdot 37^2$
$8e$	$(1,2,3,7,8,8,8,8)$	2	8	5040	$2^6 \cdot 3^3$	$3 \cdot 17^2$
$9a$	$(1,2,4,6,8,8,8,8)$	3	1	1260	$3^3 \cdot 71^2$	$3 \cdot 107^2$
$9b$	$(1,2,4,6,8,8,8,8)$	3	2	1260	$3^3 \cdot 71^2$	$3 \cdot 17^2$
$9c$	$(1,2,4,6,8,8,8,8)$	3	4	1260	$3^3 \cdot 71^2$	3
$9d$	$(1,2,4,6,8,8,8,8)$	3	6	1260	$3^3 \cdot 71^2$	3^5
$9e$	$(1,2,4,6,8,8,8,8)$	3	8	5040	$3^3 \cdot 71^2$	$3 \cdot 233^2$
$10a$	$(1,2,5,5,8,8,8,8)$	1	1	630	3^3	$3 \cdot 17^2$
$10b$	$(1,2,5,5,8,8,8,8)$	1	2	630	3^3	3
$10c$	$(1,2,5,5,8,8,8,8)$	1	5	1260	3^3	$2^6 \cdot 3$
$10d$	$(1,2,5,5,8,8,8,8)$	1	8	2520	3^3	3
$11a$	$(1,3,3,6,8,8,8,8)$	2	1	630	$2^6 \cdot 3^3$	$3 \cdot 19^2$
$11b$	$(1,3,3,6,8,8,8,8)$	2	3	1260	$2^6 \cdot 3^3$	3^3
$11c$	$(1,3,3,6,8,8,8,8)$	2	6	630	$2^6 \cdot 3^3$	3^7
$11d$	$(1,3,3,6,8,8,8,8)$	2	8	2520	$2^6 \cdot 3^3$	$3 \cdot 17^2$
$12a$	$(1,3,4,5,8,8,8,8)$	2	1	1260	$2^6 \cdot 3^3$	$3 \cdot 19^2$
$12b$	$(1,3,4,5,8,8,8,8)$	2	3	1260	$2^6 \cdot 3^3$	3^3
$12c$	$(1,3,4,5,8,8,8,8)$	2	4	1260	$2^6 \cdot 3^3$	3
$12d$	$(1,3,4,5,8,8,8,8)$	2	5	1260	$2^6 \cdot 3^3$	3
$12e$	$(1,3,4,5,8,8,8,8)$	2	8	5040	$2^6 \cdot 3^3$	$3 \cdot 17^2$
$13a$	$(1,4,4,4,8,8,8,8)$	4	1	210	$2^6 \cdot 3^9$	$3 \cdot 127^2$
$13b$	$(1,4,4,4,8,8,8,8)$	4	4	630	$2^6 \cdot 3^9$	$3 \cdot 17^2 \cdot 71^2$
$13c$	$(1,4,4,4,8,8,8,8)$	4	8	840	$2^6 \cdot 3^9$	$3 \cdot 919^2$
$14a$	$(1,7,7,7,8,8,8,8)$	4	1	210	$2^6 \cdot 3^9$	$3 \cdot 163^2$
$14b$	$(1,7,7,7,8,8,8,8)$	4	7	630	$2^6 \cdot 3^9$	$3 \cdot 919^2$
$14c$	$(1,7,7,7,8,8,8,8)$	4	8	840	$2^6 \cdot 3^9$	$3 \cdot 17^2 \cdot 71^2$
$15a$	$*(1,1,1,2,7,8,8,8)$	0	1	1260	0	3
$15b$	$*(1,1,1,2,7,8,8,8)$	0	2	420	0	3
$15c$	$*(1,1,1,2,7,8,8,8)$	0	7	420	0	3
$15d$	$*(1,1,1,2,7,8,8,8)$	0	8	1260	0	3

	a	$w(3)$	a_0	mult.	$c_1 = 1$	c_2
16a	$*(1,1,1,3,6,8,8,8)$	0	1	1260	0	3
16b	$*(1,1,1,3,6,8,8,8)$	0	3	420	0	3^3
16c	$*(1,1,1,3,6,8,8,8)$	0	6	420	0	3^3
16d	$*(1,1,1,3,6,8,8,8)$	0	8	1260	0	3
17a	$*(1,1,1,4,5,8,8,8)$	0	1	1260	0	3
17b	$*(1,1,1,4,5,8,8,8)$	0	4	420	0	3
17c	$*(1,1,1,4,5,8,8,8)$	0	5	420	0	3
17d	$*(1,1,1,4,5,8,8,8)$	0	8	1260	0	3
18a	$(1,1,2,2,6,8,8,8)$	1	1	2520	3^3	$3 \cdot 17^2$
18b	$(1,1,2,2,6,8,8,8)$	1	2	2520	3^3	$2^6 \cdot 3$
18c	$(1,1,2,2,6,8,8,8)$	1	6	1260	3^3	3^5
18d	$(1,1,2,2,6,8,8,8)$	1	8	3780	3^3	3
19a	$*(1,1,2,3,5,8,8,8)$	0	1	5040	3^3	3
19b	$*(1,1,2,3,5,8,8,8)$	0	2	2520	3^3	3
19c	$*(1,1,2,3,5,8,8,8)$	0	3	2520	3^3	3^3
19d	$*(1,1,2,3,5,8,8,8)$	0	5	2520	3^3	3
19e	$*(1,1,2,3,5,8,8,8)$	0	8	7560	3^3	3
20a	$(1,1,2,4,4,8,8,8)$	2	1	2520	3^7	$3 \cdot 17^2$
20b	$(1,1,2,4,4,8,8,8)$	2	2	1260	3^7	$3 \cdot 19^2$
20c	$(1,1,2,4,4,8,8,8)$	2	4	2520	3^7	3
20d	$(1,1,2,4,4,8,8,8)$	2	8	3780	3^7	3
21a	$(1,1,3,3,4,8,8,8)$	1	1	2520	3^5	$2^6 \cdot 3$
21b	$(1,1,3,3,4,8,8,8)$	1	3	2520	3^5	3^3
21c	$(1,1,3,3,4,8,8,8)$	1	4	1260	3^5	3
21d	$(1,1,3,3,4,8,8,8)$	1	8	3780	3^5	$3 \cdot 17^2$
22a	$(1,1,5,7,7,8,8,8)$	2	1	2520	3^7	$3 \cdot 73^2$
22b	$(1,1,5,7,7,8,8,8)$	2	5	1260	3^7	3
22c	$(1,1,5,7,7,8,8,8)$	2	7	2520	3^7	3
22d	$(1,1,5,7,7,8,8,8)$	2	8	3780	3^7	$3 \cdot 19^2$
23a	$(1,1,6,6,7,8,8,8)$	1	1	2520	3^5	3

	a	$w(3)$	a_0	mult.	$\mathbf{c_1 = 1}$	$\mathbf{c_2}$
23b	$(1,1,6,6,7,8,8,8)$	1	6	2520	3^5	3^3
23c	$(1,1,6,6,7,8,8,8)$	1	7	1260	3^5	$3 \cdot 17^2$
23d	$(1,1,6,6,7,8,8,8)$	1	8	3780	3^5	3
24a	$(1,2,2,2,5,8,8,8)$	1	1	840	3^3	3
24b	$(1,2,2,2,5,8,8,8)$	1	2	2520	3^3	3
24c	$(1,2,2,2,5,8,8,8)$	1	5	840	3^3	3
24d	$(1,2,2,2,5,8,8,8)$	1	8	2520	3^3	$2^6 \cdot 3$
25a	$(1,2,2,3,4,8,8,8)$	2	1	2520	3^3	3
25b	$(1,2,2,3,4,8,8,8)$	2	2	5040	3^3	$3 \cdot 37^2$
25c	$(1,2,2,3,4,8,8,8)$	2	3	2520	3^3	$2^6 \cdot 3^3$
25d	$(1,2,2,3,4,8,8,8)$	2	4	2520	3^3	$3 \cdot 73^2$
25e	$(1,2,2,3,4,8,8,8)$	2	8	7560	3^3	$3 \cdot 19^2$
26a	$(1,2,3,3,3,8,8,8)$	1	1	840	3^3	3
26b	$(1,2,3,3,3,8,8,8)$	1	2	840	3^3	3
26c	$(1,2,3,3,3,8,8,8)$	1	3	2520	3^3	3^3
26d	$(1,2,3,3,3,8,8,8)$	1	8	2520	3^3	3
27a	$(1,2,4,7,7,8,8,8)$	2	1	2520	3^7	$3 \cdot 37^2$
27b	$(1,2,4,7,7,8,8,8)$	2	2	2520	3^7	$3 \cdot 17^2$
27c	$(1,2,4,7,7,8,8,8)$	2	4	2520	3^7	$3 \cdot 19^2$
27d	$(1,2,4,7,7,8,8,8)$	2	7	5040	3^7	3
27e	$(1,2,4,7,7,8,8,8)$	2	8	7560	3^7	3
28a	$(1,2,5,6,7,8,8,8)$	1	1	5040	3^3	3
28b	$(1,2,5,6,7,8,8,8)$	1	2	5040	3^3	3
28c	$(1,2,5,6,7,8,8,8)$	1	5	5040	3^3	3
28d	$(1,2,5,6,7,8,8,8)$	1	6	5040	3^3	3^5
28e	$(1,2,5,6,7,8,8,8)$	1	7	5040	3^3	3
28f	$(1,2,5,6,7,8,8,8)$	1	8	15120	3^3	$2^6 \cdot 3$
29a	$(1,2,6,6,6,8,8,8)$	2	1	840	3^3	$3 \cdot 19^2$
29b	$(1,2,6,6,6,8,8,8)$	2	2	840	3^3	$3 \cdot 53^2$
29c	$(1,2,6,6,6,8,8,8)$	2	6	2520	3^3	3^5

	a	$w(3)$	a_0	mult.	$c_1 = 1$	c_2
29d	$(1,2,6,6,6,8,8,8)$	2	8	2520	3^3	3
30a	$(1,3,3,7,7,8,8,8)$	3	1	1260	3^5	$3 \cdot 17^2$
30b	$(1,3,3,7,7,8,8,8)$	3	3	2520	3^5	$3^3 \cdot 17^2$
30c	$(1,3,3,7,7,8,8,8)$	3	7	2520	3^5	$2^6 \cdot 3 \cdot 17^2$
30d	$(1,3,3,7,7,8,8,8)$	3	8	3780	3^5	$3 \cdot 107^2$
31a	$(1,3,4,6,7,8,8,8)$	2	1	5040	3^7	$3 \cdot 37^2$
31b	$(1,3,4,6,7,8,8,8)$	2	3	5040	3^7	$2^6 \cdot 3^3$
31c	$(1,3,4,6,7,8,8,8)$	2	4	5040	3^7	$3 \cdot 19^2$
31d	$(1,3,4,6,7,8,8,8)$	2	6	5040	3^7	3^3
31e	$(1,3,4,6,7,8,8,8)$	2	7	5040	3^7	3
31f	$(1,3,4,6,7,8,8,8)$	2	8	15120	3^7	3
32a	$(1,3,5,5,7,8,8,8)$	2	1	2520	3^3	$3 \cdot 17^2$
32b	$(1,3,5,5,7,8,8,8)$	2	3	2520	3^3	$2^6 \cdot 3^3$
32c	$(1,3,5,5,7,8,8,8)$	2	5	5040	3^3	$3 \cdot 19^2$
32d	$(1,3,5,5,7,8,8,8)$	2	7	2520	3^3	$3 \cdot 73^2$
32e	$(1,3,5,5,7,8,8,8)$	2	8	7560	3^3	$3 \cdot 37^2$
33a	$(1,3,5,6,6,8,8,8)$	1	1	2520	3^3	3
33b	$(1,3,5,6,6,8,8,8)$	1	3	2520	3^3	3^3
33c	$(1,3,5,6,6,8,8,8)$	1	5	2520	3^3	3
33d	$(1,3,5,6,6,8,8,8)$	1	6	5040	3^3	3^5
33e	$(1,3,5,6,6,8,8,8)$	1	8	7560	3^3	$2^6 \cdot 3$
34a	$(1,4,4,5,7,8,8,8)$	2	1	2520	3^7	$3 \cdot 37^2$
34b	$(1,4,4,5,7,8,8,8)$	2	4	5040	3^7	$3 \cdot 19^2$
34c	$(1,4,4,5,7,8,8,8)$	2	5	2520	3^7	$3 \cdot 73^2$
34d	$(1,4,4,5,7,8,8,8)$	2	7	2520	3^7	3
34e	$(1,4,4,5,7,8,8,8)$	2	8	7560	3^7	3
35a	$(1,4,4,6,6,8,8,8)$	3	1	1260	3^5	$3 \cdot 179^2$
35b	$(1,4,4,6,6,8,8,8)$	3	4	2520	3^5	$3 \cdot 107^2$
35c	$(1,4,4,6,6,8,8,8)$	3	6	2520	3^5	$3^3 \cdot 17^2$
35d	$(1,4,4,6,6,8,8,8)$	3	8	3780	3^5	$2^6 \cdot 3 \cdot 17^2$

	a	$w(3)$	a_0	mult.	$c_1 = 1$	c_2
36a	$(1,4,5,5,6,8,8,8)$	1	1	2520	3^3	3
36b	$(1,4,5,5,6,8,8,8)$	1	4	2520	3^3	$3 \cdot 17^2$
36c	$(1,4,5,5,6,8,8,8)$	1	5	5040	3^3	3
36d	$(1,4,5,5,6,8,8,8)$	1	6	2520	3^3	3^5
36e	$(1,4,5,5,6,8,8,8)$	1	8	7560	3^3	$2^6 \cdot 3$
37a	$(1,5,5,5,5,8,8,8)$	3	1	210	$3^3 \cdot 17^2$	$3 \cdot 107^2$
37b	$(1,5,5,5,5,8,8,8)$	3	5	840	$3^3 \cdot 17^2$	$3 \cdot 17^2$
37c	$(1,5,5,5,5,8,8,8)$	3	8	630	$3^3 \cdot 17^2$	$2^6 \cdot 3 \cdot 17^2$
38a	$(1,1,2,2,2,3,8,8)$	2	1	2520	$2^6 \cdot 3^3$	3
38b	$(1,1,2,2,2,3,8,8)$	2	2	3780	$2^6 \cdot 3^3$	$3 \cdot 17^2$
38c	$(1,1,2,2,2,3,8,8)$	2	3	1260	$2^6 \cdot 3^3$	3^3
38d	$(1,1,2,2,2,3,8,8)$	2	8	2520	$2^6 \cdot 3^3$	3
39a	$*(1,1,2,2,7,7,8,8)$	0	1	1890	0	3
39b	$*(1,1,2,2,7,7,8,8)$	0	2	1890	0	3
39c	$*(1,1,2,2,7,7,8,8)$	0	7	1890	0	3
39d	$*(1,1,2,2,7,7,8,8)$	0	8	1890	0	3
40a	$*(1,1,2,3,6,7,8,8)$	0	1	7560	0	3
40b	$*(1,1,2,3,6,7,8,8)$	0	2	3780	0	3
40c	$*(1,1,2,3,6,7,8,8)$	0	3	3780	0	3^3
40d	$*(1,1,2,3,6,7,8,8)$	0	6	3780	0	3^3
40e	$*(1,1,2,3,6,7,8,8)$	0	7	3780	0	3
40f	$*(1,1,2,3,6,7,8,8)$	0	8	7560	0	3
41a	$*(1,1,2,4,5,7,8,8)$	0	1	7560	0	3
41b	$*(1,1,2,4,5,7,8,8)$	0	2	3780	0	3
41c	$*(1,1,2,4,5,7,8,8)$	0	4	3780	0	3
41d	$*(1,1,2,4,5,7,8,8)$	0	5	3780	0	3
41e	$*(1,1,2,4,5,7,8,8)$	0	7	3780	0	3
41f	$*(1,1,2,4,5,7,8,8)$	0	8	7560	0	3
42a	$(1,1,2,4,6,6,8,8)$	1	1	7560	3^5	3
42b	$(1,1,2,4,6,6,8,8)$	1	2	3780	3^5	3

	a	$w(3)$	a_0	mult.	$c_1 = 1$	c_2
42c	$(1,1,2,4,6,6,8,8)$	1	4	3780	3^5	$3 \cdot 17^2$
42d	$(1,1,2,4,6,6,8,8)$	1	6	7560	3^5	3^3
42e	$(1,1,2,4,6,6,8,8)$	1	8	7560	3^5	3
43a	$(1,1,2,5,5,6,8,8)$	1	1	7560	3^3	3
43b	$(1,1,2,5,5,6,8,8)$	1	2	3780	3^3	3
43c	$(1,1,2,5,5,6,8,8)$	1	5	7560	3^3	$2^6 \cdot 3$
43d	$(1,1,2,5,5,6,8,8)$	1	6	3780	3^3	3^5
43e	$(1,1,2,5,5,6,8,8)$	1	8	7560	3^3	3
44a	$*(1,1,3,3,6,6,8,8)$	0	1	1890	0	3
44b	$*(1,1,3,3,6,6,8,8)$	0	3	1890	0	3^3
44c	$*(1,1,3,3,6,6,8,8)$	0	6	1890	0	3^3
44d	$*(1,1,3,3,6,6,8,8)$	0	8	1890	0	3
45a	$*(1,1,3,4,5,6,8,8)$	0	1	7560	0	3
45b	$*(1,1,3,4,5,6,8,8)$	0	3	3780	0	3^3
45c	$*(1,1,3,4,5,6,8,8)$	0	4	3780	0	3
45d	$*(1,1,3,4,5,6,8,8)$	0	5	3780	0	3
45e	$*(1,1,3,4,5,6,8,8)$	0	6	3780	0	3^3
45f	$*(1,1,3,4,5,6,8,8)$	0	8	7560	0	3
46a	$(1,1,3,5,5,5,8,8)$	2	1	2520	$2^6 \cdot 3^3$	$3 \cdot 37^2$
46b	$(1,1,3,5,5,5,8,8)$	2	3	1260	$2^6 \cdot 3^3$	3^3
46c	$(1,1,3,5,5,5,8,8)$	2	5	3780	$2^6 \cdot 3^3$	$3 \cdot 17^2$
46d	$(1,1,3,5,5,5,8,8)$	2	8	2520	$2^6 \cdot 3^3$	$3 \cdot 73^2$
47a	$(1,2,2,2,2,2,8,8)$	3	1	126	$3^3 \cdot 109^2$	$3 \cdot 5^6$
47b	$(1,2,2,2,2,2,8,8)$	3	2	630	$3^3 \cdot 109^2$	$2^6 \cdot 3$
47c	$(1,2,2,2,2,2,8,8)$	3	8	252	$3^3 \cdot 109^2$	$3 \cdot 53^2$
48a	$*(1,2,2,3,5,7,8,8)$	0	1	7560	3^3	3
48b	$*(1,2,2,3,5,7,8,8)$	0	2	15120	3^3	3
48c	$*(1,2,2,3,5,7,8,8)$	0	3	7560	3^3	3^3
48d	$*(1,2,2,3,5,7,8,8)$	0	5	7560	3^3	3
48e	$*(1,2,2,3,5,7,8,8)$	0	7	7560	3^3	3

	a	$w(3)$	a_0	mult.	$\mathbf{c_1 = 1}$	$\mathbf{c_2}$
48f	$*(1,2,2,3,5,7,8,8)$	0	8	15120	3^3	3
49a	$(1,2,2,3,6,6,8,8)$	1	1	3780	3^3	$3 \cdot 17^2$
49b	$(1,2,2,3,6,6,8,8)$	1	2	7560	3^3	$2^6 \cdot 3$
49c	$(1,2,2,3,6,6,8,8)$	1	3	3780	3^3	3^3
49d	$(1,2,2,3,6,6,8,8)$	1	6	7560	3^3	3^5
49e	$(1,2,2,3,6,6,8,8)$	1	8	7560	3^3	3
50a	$(1,2,2,4,4,7,8,8)$	2	1	3780	3^7	$3 \cdot 17^2$
50b	$(1,2,2,4,4,7,8,8)$	2	2	7560	3^7	$3 \cdot 19^2$
50c	$(1,2,2,4,4,7,8,8)$	2	4	7560	3^7	3
50d	$(1,2,2,4,4,7,8,8)$	2	7	3780	3^7	$3 \cdot 73^2$
50e	$(1,2,2,4,4,7,8,8)$	2	8	7560	3^7	3
51a	$(1,2,2,4,5,6,8,8)$	1	1	7560	3^3	$3 \cdot 17^2$
51b	$(1,2,2,4,5,6,8,8)$	1	2	15120	3^3	$2^6 \cdot 3$
51c	$(1,2,2,4,5,6,8,8)$	1	4	7560	3^3	3
51d	$(1,2,2,4,5,6,8,8)$	1	5	7560	3^3	3
51e	$(1,2,2,4,5,6,8,8)$	1	6	7560	3^3	3^5
51f	$(1,2,2,4,5,6,8,8)$	1	8	15120	3^3	3
52a	$(1,2,2,5,5,5,8,8)$	1	1	1260	3^3	3
52b	$(1,2,2,5,5,5,8,8)$	1	2	2520	3^3	$2^6 \cdot 3$
52c	$(1,2,2,5,5,5,8,8)$	1	5	3780	3^3	3
52d	$(1,2,2,5,5,5,8,8)$	1	8	2520	3^3	3
53a	$(1,2,3,3,4,7,8,8)$	1	1	7560	3^5	$2^6 \cdot 3$
53b	$(1,2,3,3,4,7,8,8)$	1	2	7560	3^5	3
53c	$(1,2,3,3,4,7,8,8)$	1	3	15120	3^5	3^3
53d	$(1,2,3,3,4,7,8,8)$	1	4	7560	3^5	3
53e	$(1,2,3,3,4,7,8,8)$	1	7	7560	3^5	3
53f	$(1,2,3,3,4,7,8,8)$	1	8	15120	3^5	$3 \cdot 17^2$
54a	$*(1,2,3,3,5,6,8,8)$	0	1	7560	3^3	3
54b	$*(1,2,3,3,5,6,8,8)$	0	2	7560	3^3	3
54c	$*(1,2,3,3,5,6,8,8)$	0	3	15120	3^3	3^3

	a	$w(3)$	a_0	mult.	$c_1 = 1$	c_2
54d	$*(1,2,3,3,5,6,8,8)$	0	5	7560	3^3	3
54e	$*(1,2,3,3,5,6,8,8)$	0	6	7560	3^3	0
54f	$*(1,2,3,3,5,6,8,8)$	0	8	15120	3^3	3
55a	$(1,2,3,4,4,6,8,8)$	2	1	7560	3^7	$3 \cdot 17^2$
55b	$(1,2,3,4,4,6,8,8)$	2	2	7560	3^7	$3 \cdot 19^2$
55c	$(1,2,3,4,4,6,8,8)$	2	3	7560	3^7	3^3
55d	$(1,2,3,4,4,6,8,8)$	2	4	15120	3^7	3
55e	$(1,2,3,4,4,6,8,8)$	2	6	7560	3^7	$2^6 \cdot 3^3$
55f	$(1,2,3,4,4,6,8,8)$	2	8	15120	3^7	3
56a	$(1,2,6,6,7,7,8,8)$	1	1	3780	3^5	3
56b	$(1,2,6,6,7,7,8,8)$	1	2	3780	3^5	$2^6 \cdot 3$
56c	$(1,2,6,6,7,7,8,8)$	1	6	7560	3^5	3^3
56d	$(1,2,6,6,7,7,8,8)$	1	7	7560	3^5	$3 \cdot 17^2$
56e	$(1,2,6,6,7,7,8,8)$	1	8	7560	3^5	3
57a	$(1,3,3,3,3,7,8,8)$	2	1	630	3^5	$3 \cdot 53^2$
57b	$(1,3,3,3,3,7,8,8)$	2	3	2520	3^5	$3^3 \cdot 17^2$
57c	$(1,3,3,3,3,7,8,8)$	2	7	630	3^5	$3 \cdot 37^2$
57d	$(1,3,3,3,3,7,8,8)$	2	8	1260	3^5	$3 \cdot 19^2$
58a	$(1,3,3,3,4,6,8,8)$	1	1	2520	3^5	$2^6 \cdot 3$
58b	$(1,3,3,3,4,6,8,8)$	1	3	7560	3^5	3^3
58c	$(1,3,3,3,4,6,8,8)$	1	4	2520	3^5	3
58d	$(1,3,3,3,4,6,8,8)$	1	6	2520	3^5	3^3
58e	$(1,3,3,3,4,6,8,8)$	1	8	5040	3^5	$3 \cdot 17^2$
59a	$(1,3,3,3,5,5,8,8)$	1	1	1260	3^3	$2^6 \cdot 3$
59b	$(1,3,3,3,5,5,8,8)$	1	3	3780	3^3	3^3
59c	$(1,3,3,3,5,5,8,8)$	1	5	2520	3^3	3
59d	$(1,3,3,3,5,5,8,8)$	1	8	2520	3^3	3
60a	$(1,3,4,4,4,4,8,8)$	3	1	630	3^3	$2^6 \cdot 3$
60b	$(1,3,4,4,4,4,8,8)$	3	3	630	3^3	$3^5 \cdot 17^2$
60c	$(1,3,4,4,4,4,8,8)$	3	4	2520	3^3	$3 \cdot 53^2$

	a	$w(3)$	a_0	mult.	$c_1 = 1$	c_2
60d	$(1,3,4,4,4,4,8,8)$	3	8	1260	3^3	$3 \cdot 5^6$
61a	$(1,3,4,7,7,7,8,8)$	3	1	2520	$3^3 \cdot 71^2$	$2^6 \cdot 3 \cdot 17^2$
61b	$(1,3,4,7,7,7,8,8)$	3	3	2520	$3^3 \cdot 71^2$	3^5
61c	$(1,3,4,7,7,7,8,8)$	3	4	2520	$3^3 \cdot 71^2$	$3 \cdot 17^2$
61d	$(1,3,4,7,7,7,8,8)$	3	7	7560	$3^3 \cdot 71^2$	$3 \cdot 233^2$
61e	$(1,3,4,7,7,7,8,8)$	3	8	5040	$3^3 \cdot 71^2$	3
62a	$(1,3,5,6,7,7,8,8)$	2	1	7560	3^7	$3 \cdot 73^2$
62b	$(1,3,5,6,7,7,8,8)$	2	3	7560	3^7	3^3
62c	$(1,3,5,6,7,7,8,8)$	2	5	7560	3^7	3
62d	$(1,3,5,6,7,7,8,8)$	2	6	7560	3^7	$2^6 \cdot 3^3$
62e	$(1,3,5,6,7,7,8,8)$	2	7	15120	3^7	3
62f	$(1,3,5,6,7,7,8,8)$	2	8	15120	3^7	$3 \cdot 19^2$
63a	$(1,3,6,6,6,7,8,8)$	1	1	2520	3^5	3
63b	$(1,3,6,6,6,7,8,8)$	1	3	2520	3^5	3^3
63c	$(1,3,6,6,6,7,8,8)$	1	6	7560	3^5	3^3
63d	$(1,3,6,6,6,7,8,8)$	1	7	2520	3^5	$3 \cdot 17^2$
63e	$(1,3,6,6,6,7,8,8)$	1	8	5040	3^5	3
64a	$(1,4,4,6,7,7,8,8)$	2	1	3780	3^3	3
64b	$(1,4,4,6,7,7,8,8)$	2	4	7560	3^3	$3 \cdot 37^2$
64c	$(1,4,4,6,7,7,8,8)$	2	6	3780	3^3	$2^6 \cdot 3^3$
64d	$(1,4,4,6,7,7,8,8)$	2	7	7560	3^3	$3 \cdot 19^2$
64e	$(1,4,4,6,7,7,8,8)$	2	8	7560	3^3	$3 \cdot 73^2$
65a	$(1,4,5,6,6,7,8,8)$	1	1	7560	3^5	3
65b	$(1,4,5,6,6,7,8,8)$	1	4	7560	3^5	3
65c	$(1,4,5,6,6,7,8,8)$	1	5	7560	3^5	3
65d	$(1,4,5,6,6,7,8,8)$	1	6	15120	3^5	3^3
65e	$(1,4,5,6,6,7,8,8)$	1	7	7560	3^5	$3 \cdot 17^2$
65f	$(1,4,5,6,6,7,8,8)$	1	8	15120	3^5	3
66a	$(1,4,6,6,6,6,8,8)$	2	1	630	3^5	$3 \cdot 37^2$
66b	$(1,4,6,6,6,6,8,8)$	2	4	630	3^5	$3 \cdot 19^2$

	a	$w(3)$	a_0	mult.	$c_1 = 1$	c_2
66c	$(1,4,6,6,6,6,8,8)$	2	6	2520	3^5	$3^3 \cdot 17^2$
66d	$(1,4,6,6,6,6,8,8)$	2	8	1260	3^5	$3 \cdot 37^2$
67a	$(1,5,5,5,6,7,8,8)$	3	1	2520	$3^3 \cdot 71^2$	$3 \cdot 179^2$
67b	$(1,5,5,5,6,7,8,8)$	3	5	7560	$3^3 \cdot 71^2$	$3 \cdot 233^2$
67c	$(1,5,5,5,6,7,8,8)$	3	6	2520	$3^3 \cdot 71^2$	3^5
67d	$(1,5,5,5,6,7,8,8)$	3	7	2520	$3^3 \cdot 71^2$	3
67e	$(1,5,5,5,6,7,8,8)$	3	8	5040	$3^3 \cdot 71^2$	$3 \cdot 17^2$
68a	$(1,5,5,6,6,6,8,8)$	2	1	1260	3^3	$3 \cdot 37^2$
68b	$(1,5,5,6,6,6,8,8)$	2	5	2520	3^3	3
68c	$(1,5,5,6,6,6,8,8)$	2	6	3780	3^3	3^5
68d	$(1,5,5,6,6,6,8,8)$	2	8	2520	3^3	$3 \cdot 53^2$
69a	$(1,1,1,1,1,1,1,2)$	5	1	42	$3^3 \cdot 233^2$	$3 \cdot 2953^2$
69b	$(1,1,1,1,1,1,1,2)$	5	2	6	$3^3 \cdot 233^2$	$3 \cdot 2393^2$
70a	$(1,1,1,1,1,1,5,7)$	5	1	252	$3^3 \cdot 163^2$	$3 \cdot 863^2$
70b	$(1,1,1,1,1,1,5,7)$	5	5	42	$3^3 \cdot 163^2$	$3 \cdot 6029^2$
70c	$(1,1,1,1,1,1,5,7)$	5	7	42	$3^3 \cdot 163^2$	$3 \cdot 17^2 \cdot 127^2$
71a	$(1,1,1,1,1,1,6,6)$	4	1	126	$2^{12} \cdot 3^3$	$3 \cdot 37^2$
71b	$(1,1,1,1,1,1,6,6)$	4	6	42	$2^{12} \cdot 3^3$	$3^7 \cdot 17^2$
72a	$(1,1,1,1,1,2,4,7)$	3	1	1260	$3^3 \cdot 109^2$	$2^6 \cdot 3$
72b	$(1,1,1,1,1,2,4,7)$	3	2	252	$3^3 \cdot 109^2$	$3 \cdot 89^2$
72c	$(1,1,1,1,1,2,4,7)$	3	4	252	$3^3 \cdot 109^2$	$3 \cdot 53^2$
72d	$(1,1,1,1,1,2,4,7)$	3	7	252	$3^3 \cdot 109^2$	$3 \cdot 53^2$
73a	$(1,1,1,1,1,2,5,6)$	4	1	1260	$3^3 \cdot 89^2$	$3 \cdot 37^4$
73b	$(1,1,1,1,1,2,5,6)$	4	2	252	$3^3 \cdot 89^2$	$3 \cdot 73^2$
73c	$(1,1,1,1,1,2,5,6)$	4	5	252	$3^3 \cdot 89^2$	$3 \cdot 109^2$
73d	$(1,1,1,1,1,2,5,6)$	4	6	252	$3^3 \cdot 89^2$	$2^{12} \cdot 3^3$
74a	$(1,1,1,1,1,3,3,7)$	4	1	630	$3^3 \cdot 17^2$	$3 \cdot 359^2$
74b	$(1,1,1,1,1,3,3,7)$	4	3	252	$3^3 \cdot 17^2$	$3^3 \cdot 271^2$
74c	$(1,1,1,1,1,3,3,7)$	4	7	126	$3^3 \cdot 17^2$	$3 \cdot 307^2$
75a	$(1,1,1,1,1,3,4,6)$	3	1	1260	$3^3 \cdot 109^2$	$2^6 \cdot 3$

	a	$w(3)$	a_0	mult.	$c_1 = 1$	c_2
75b	$(1,1,1,1,1,3,4,6)$	3	3	252	$3^3 \cdot 109^2$	3^3
75c	$(1,1,1,1,1,3,4,6)$	3	4	252	$3^3 \cdot 109^2$	$3 \cdot 53^2$
75d	$(1,1,1,1,1,3,4,6)$	3	6	252	$3^3 \cdot 109^2$	$3^5 \cdot 17^2$
76a	$(1,1,1,1,1,3,5,5)$	5	1	630	$3^3 \cdot 379^2$	$2^6 \cdot 3 \cdot 107^2$
76b	$(1,1,1,1,1,3,5,5)$	5	3	126	$3^3 \cdot 379^2$	$3^5 \cdot 433^2$
76c	$(1,1,1,1,1,3,5,5)$	5	5	252	$3^3 \cdot 379^2$	$3 \cdot 433^2$
77a	$(1,1,1,1,2,2,4,6)$	4	1	2520	$3^3 \cdot 179^2$	$3 \cdot 127^2$
77b	$(1,1,1,1,2,2,4,6)$	4	2	1260	$3^3 \cdot 179^2$	$3 \cdot 163^2$
77c	$(1,1,1,1,2,2,4,6)$	4	4	630	$3^3 \cdot 179^2$	$3 \cdot 17^2 \cdot 71^2$
77d	$(1,1,1,1,2,2,4,6)$	4	6	630	$3^3 \cdot 179^2$	$3^3 \cdot 37^2$
78a	$(1,1,1,1,2,2,5,5)$	4	1	1260	$3^7 \cdot 17^2$	$3 \cdot 17^2 \cdot 19^2$
78b	$(1,1,1,1,2,2,5,5)$	4	2	630	$3^7 \cdot 17^2$	$3 \cdot 109^2$
78c	$(1,1,1,1,2,2,5,5)$	4	5	630	$3^7 \cdot 17^2$	$3 \cdot 17^2 \cdot 53^2$
79a	$(1,1,1,1,2,3,3,6)$	3	1	2520	3^3	$3 \cdot 53^2$
79b	$(1,1,1,1,2,3,3,6)$	3	2	630	3^3	$3 \cdot 5^6$
79c	$(1,1,1,1,2,3,3,6)$	3	3	1260	3^3	$3^5 \cdot 17^2$
79d	$(1,1,1,1,2,3,3,6)$	3	6	630	3^3	$3^3 \cdot 109^2$
80a	$(1,1,1,1,2,3,4,5)$	3	1	5040	3^3	$3 \cdot 53^2$
80b	$(1,1,1,1,2,3,4,5)$	3	2	1260	3^3	$3 \cdot 5^6$
80c	$(1,1,1,1,2,3,4,5)$	3	3	1260	3^3	$3^5 \cdot 17^2$
80d	$(1,1,1,1,2,3,4,5)$	3	4	1260	3^3	$3 \cdot 53^2$
80e	$(1,1,1,1,2,3,4,5)$	3	5	1260	3^3	$3 \cdot 71^2$
81a	$(1,1,1,1,2,4,4,4)$	3	1	840	3^3	$3 \cdot 269^2$
81b	$(1,1,1,1,2,4,4,4)$	3	2	210	3^3	$3 \cdot 17^2$
81c	$(1,1,1,1,2,4,4,4)$	3	4	630	3^3	$3 \cdot 163^2$
82a	$(1,1,1,1,3,3,3,5)$	4	1	840	$3^3 \cdot 271^2$	$3 \cdot 233^2$
82b	$(1,1,1,1,3,3,3,5)$	4	3	630	$3^3 \cdot 271^2$	$3^5 \cdot 37^2$
82c	$(1,1,1,1,3,3,3,5)$	4	5	210	$3^3 \cdot 271^2$	$3 \cdot 37^4$
83a	$(1,1,1,1,3,3,4,4)$	2	1	1260	3^3	$3 \cdot 17^2$
83b	$(1,1,1,1,3,3,4,4)$	2	3	630	3^3	$3^3 \cdot 19^2$

	a	$w(3)$	a_0	mult.	$c_1 = 1$	c_2
83c	$(1,1,1,1,3,3,4,4)$	2	4	630	3^3	$3 \cdot 19^2$
84a	$(1,1,1,1,3,6,7,7)$	3	1	2520	$3^3 \cdot 17^2$	$3 \cdot 17^2$
84b	$(1,1,1,1,3,6,7,7)$	3	3	630	$3^3 \cdot 17^2$	$3^3 \cdot 71^2$
84c	$(1,1,1,1,3,6,7,7)$	3	6	630	$3^3 \cdot 17^2$	3^5
84d	$(1,1,1,1,3,6,7,7)$	3	7	1260	$3^3 \cdot 17^2$	$2^6 \cdot 3 \cdot 17^2$
85a	$(1,1,1,1,4,5,7,7)$	3	1	2520	$3^3 \cdot 17^2$	$3 \cdot 17^2$
85b	$(1,1,1,1,4,5,7,7)$	3	4	630	$3^3 \cdot 17^2$	$3 \cdot 233^2$
85c	$(1,1,1,1,4,5,7,7)$	3	5	630	$3^3 \cdot 17^2$	$3 \cdot 179^2$
85d	$(1,1,1,1,4,5,7,7)$	3	7	1260	$3^3 \cdot 17^2$	$2^6 \cdot 3 \cdot 17^2$
86a	$(1,1,1,1,4,6,6,7)$	2	1	2520	3^3	$3 \cdot 73^2$
86b	$(1,1,1,1,4,6,6,7)$	2	4	630	3^3	$3 \cdot 17^2$
86c	$(1,1,1,1,4,6,6,7)$	2	6	1260	3^3	3^7
86d	$(1,1,1,1,4,6,6,7)$	2	7	630	3^3	3
87a	$(1,1,1,1,5,5,6,7)$	4	1	2520	$3^3 \cdot 37^2$	3
87b	$(1,1,1,1,5,5,6,7)$	4	5	1260	$3^3 \cdot 37^2$	$3 \cdot 251^2$
87c	$(1,1,1,1,5,5,6,7)$	4	6	630	$3^3 \cdot 37^2$	$3^3 \cdot 179^2$
87d	$(1,1,1,1,5,5,6,7)$	4	7	630	$3^3 \cdot 37^2$	$3 \cdot 163^2$
88a	$(1,1,1,1,5,6,6,6)$	3	1	840	3^3	$3 \cdot 17^2$
88b	$(1,1,1,1,5,6,6,6)$	3	5	210	3^3	$3 \cdot 269^2$
88c	$(1,1,1,1,5,6,6,6)$	3	6	630	3^3	$3^3 \cdot 53^2$
89a	$(1,1,1,2,2,2,3,6)$	4	1	2520	$2^6 \cdot 3^9$	$3 \cdot 17^2 \cdot 71^2$
89b	$(1,1,1,2,2,2,3,6)$	4	2	2520	$2^6 \cdot 3^9$	$3 \cdot 919^2$
89c	$(1,1,1,2,2,2,3,6)$	4	3	840	$2^6 \cdot 3^9$	$3^3 \cdot 37^2$
89d	$(1,1,1,2,2,2,3,6)$	4	6	840	$2^6 \cdot 3^9$	$3^3 \cdot 179^2$
90a	$(1,1,1,2,2,2,4,5)$	4	1	2520	$2^6 \cdot 3^9$	$3 \cdot 17^2 \cdot 71^2$
90b	$(1,1,1,2,2,2,4,5)$	4	2	2520	$2^6 \cdot 3^9$	$3 \cdot 919^2$
90c	$(1,1,1,2,2,2,4,5)$	4	4	840	$2^6 \cdot 3^9$	$3 \cdot 251^2$
90d	$(1,1,1,2,2,2,4,5)$	4	5	840	$2^6 \cdot 3^9$	3
91a	$(1,1,1,2,2,3,3,5)$	3	1	3780	$3^5 \cdot 17^2$	$3 \cdot 53^2$
91b	$(1,1,1,2,2,3,3,5)$	3	2	2520	$3^5 \cdot 17^2$	$3 \cdot 71^2$

	a	$w(3)$	a_0	mult.	$\mathbf{c_1 = 1}$	$\mathbf{c_2}$
91c	$(1,1,1,2,2,3,3,5)$	3	3	2520	$3^5 \cdot 17^2$	$3^3 \cdot 109^2$
91d	$(1,1,1,2,2,3,3,5)$	3	5	1260	$3^5 \cdot 17^2$	$3 \cdot 89^2$
92a	$(1,1,1,2,2,3,4,4)$	3	1	3780	$3^3 \cdot 53^2$	$3 \cdot 163^2$
92b	$(1,1,1,2,2,3,4,4)$	3	2	2520	$3^3 \cdot 53^2$	$3 \cdot 89^2$
92c	$(1,1,1,2,2,3,4,4)$	3	3	1260	$3^3 \cdot 53^2$	3^5
92d	$(1,1,1,2,2,3,4,4)$	3	4	2520	$3^3 \cdot 53^2$	$2^6 \cdot 3$
93a	$(1,1,1,2,3,3,3,4)$	2	1	2520	$3^3 \cdot 19^2$	$3 \cdot 19^2$
93b	$(1,1,1,2,3,3,3,4)$	2	2	840	$3^3 \cdot 19^2$	$3 \cdot 17^2$
93c	$(1,1,1,2,3,3,3,4)$	2	3	2520	$3^3 \cdot 19^2$	3^5
93d	$(1,1,1,2,3,3,3,4)$	2	4	840	$3^3 \cdot 19^2$	$3 \cdot 17^2$
94a	$(1,1,1,2,3,6,6,7)$	2	1	7560	$2^6 \cdot 3^3$	$3 \cdot 17^2$
94b	$(1,1,1,2,3,6,6,7)$	2	2	2520	$2^6 \cdot 3^3$	$3 \cdot 37^2$
94c	$(1,1,1,2,3,6,6,7)$	2	3	2520	$2^6 \cdot 3^3$	3^7
94d	$(1,1,1,2,3,6,6,7)$	2	6	5040	$2^6 \cdot 3^3$	3^3
94e	$(1,1,1,2,3,6,6,7)$	2	7	2520	$2^6 \cdot 3^3$	$3 \cdot 73^2$
95a	$(1,1,1,2,4,5,6,7)$	2	1	15120	$2^6 \cdot 3^3$	$3 \cdot 17^2$
95b	$(1,1,1,2,4,5,6,7)$	2	2	5040	$2^6 \cdot 3^3$	$3 \cdot 37^2$
95c	$(1,1,1,2,4,5,6,7)$	2	4	5040	$2^6 \cdot 3^3$	3
95d	$(1,1,1,2,4,5,6,7)$	2	5	5040	$2^6 \cdot 3^3$	3
95e	$(1,1,1,2,4,5,6,7)$	2	6	5040	$2^6 \cdot 3^3$	3^3
95f	$(1,1,1,2,4,5,6,7)$	2	7	5040	$2^6 \cdot 3^3$	$3 \cdot 73^2$
96a	$(1,1,1,2,4,6,6,6)$	3	1	2520	$3^3 \cdot 17^2$	$3 \cdot 179^2$
96b	$(1,1,1,2,4,6,6,6)$	3	2	840	$3^3 \cdot 17^2$	$3 \cdot 17^2$
96c	$(1,1,1,2,4,6,6,6)$	3	4	840	$3^3 \cdot 17^2$	$3 \cdot 107^2$
96d	$(1,1,1,2,4,6,6,6)$	3	6	2520	$3^3 \cdot 17^2$	$3^3 \cdot 71^2$
97a	$(1,1,1,2,5,5,6,6)$	3	1	3780	3^5	$3 \cdot 179^2$
97b	$(1,1,1,2,5,5,6,6)$	3	2	1260	3^5	$3 \cdot 269^2$
97c	$(1,1,1,2,5,5,6,6)$	3	5	2520	3^5	$2^6 \cdot 3$
97d	$(1,1,1,2,5,5,6,6)$	3	6	2520	3^5	3^3
98a	$(1,1,1,3,3,3,3,3)$	3	1	126	$3^3 \cdot 17^2$	$3 \cdot 307^2$

	a	$w(3)$	a_0	mult.	$c_1 = 1$	c_2
98b	$(1,1,1,3,3,3,3,3)$	3	3	210	$3^3 \cdot 17^2$	$2^6 \cdot 3^7$
99a	$(1,1,1,3,3,4,7,7)$	2	1	3780	$3^3 \cdot 17^2$	$3 \cdot 53^2$
99b	$(1,1,1,3,3,4,7,7)$	2	3	2520	$3^3 \cdot 17^2$	3^3
99c	$(1,1,1,3,3,4,7,7)$	2	4	1260	$3^3 \cdot 17^2$	3
99d	$(1,1,1,3,3,4,7,7)$	2	7	2520	$3^3 \cdot 17^2$	$3 \cdot 37^2$
100a	$(1,1,1,3,3,5,6,7)$	3	1	7560	$3^3 \cdot 71^2$	$3 \cdot 233^2$
100b	$(1,1,1,3,3,5,6,7)$	3	3	5040	$3^3 \cdot 71^2$	3^5
100c	$(1,1,1,3,3,5,6,7)$	3	5	2520	$3^3 \cdot 71^2$	3
100d	$(1,1,1,3,3,5,6,7)$	3	6	2520	$3^3 \cdot 71^2$	$3^3 \cdot 17^2$
100e	$(1,1,1,3,3,5,6,7)$	3	7	2520	$3^3 \cdot 71^2$	$3 \cdot 17^2$
101a	$(1,1,1,3,3,6,6,6)$	2	1	1260	$2^6 \cdot 3^3$	$3 \cdot 17^2$
101b	$(1,1,1,3,3,6,6,6)$	2	3	840	$2^6 \cdot 3^3$	3^7
101c	$(1,1,1,3,3,6,6,6)$	2	6	1260	$2^6 \cdot 3^3$	3^3
102a	$(1,1,1,3,4,4,6,7)$	1	1	7560	3^3	3
102b	$(1,1,1,3,4,4,6,7)$	1	3	2520	3^3	3^3
102c	$(1,1,1,3,4,4,6,7)$	1	4	5040	3^3	$2^6 \cdot 3$
102d	$(1,1,1,3,4,4,6,7)$	1	6	2520	3^3	3^5
102e	$(1,1,1,3,4,4,6,7)$	1	7	2520	3^3	3
103a	$(1,1,1,3,4,5,6,6)$	2	1	7560	$2^6 \cdot 3^3$	$3 \cdot 17^2$
103b	$(1,1,1,3,4,5,6,6)$	2	3	2520	$2^6 \cdot 3^3$	3^7
103c	$(1,1,1,3,4,5,6,6)$	2	4	2520	$2^6 \cdot 3^3$	3
103d	$(1,1,1,3,4,5,6,6)$	2	5	2520	$2^6 \cdot 3^3$	3
103e	$(1,1,1,3,4,5,6,6)$	2	6	5040	$2^6 \cdot 3^3$	3^3
104a	$(1,1,1,4,4,4,6,6)$	2	1	1260	$2^6 \cdot 3^3$	3
104b	$(1,1,1,4,4,4,6,6)$	2	4	1260	$2^6 \cdot 3^3$	$3 \cdot 37^2$
104c	$(1,1,1,4,4,4,6,6)$	2	6	840	$2^6 \cdot 3^3$	3^7
105a	$(1,1,2,2,3,3,3,3)$	2	1	630	3^5	$3 \cdot 17^2$
105b	$(1,1,2,2,3,3,3,3)$	2	2	630	3^5	3
105c	$(1,1,2,2,3,3,3,3)$	2	3	1260	3^5	3^3
106a	$(1,1,2,2,3,6,6,6)$	3	1	2520	3^5	$3 \cdot 107^2$

	a	$w(3)$	a_0	mult.	$\mathbf{c_1 = 1}$	$\mathbf{c_2}$
106b	$(1,1,2,2,3,6,6,6)$	3	2	2520	3^5	$2^6 \cdot 3 \cdot 17^2$
106c	$(1,1,2,2,3,6,6,6)$	3	3	1260	3^5	$3^3 \cdot 71^2$
106d	$(1,1,2,2,3,6,6,6)$	3	6	3780	3^5	$3^3 \cdot 17^2$
107a	$(1,1,2,2,4,5,6,6)$	3	1	7560	3^5	$3 \cdot 107^2$
107b	$(1,1,2,2,4,5,6,6)$	3	2	7560	3^5	$2^6 \cdot 3 \cdot 17^2$
107c	$(1,1,2,2,4,5,6,6)$	3	4	3780	3^5	3
107d	$(1,1,2,2,4,5,6,6)$	3	5	3780	3^5	$3 \cdot 233^2$
107e	$(1,1,2,2,4,5,6,6)$	3	6	7560	3^5	$3^3 \cdot 17^2$
108a	$(1,1,2,3,3,4,6,7)$	1	1	15120	3^3	$2^6 \cdot 3$
108b	$(1,1,2,3,3,4,6,7)$	1	2	7560	3^3	3
108c	$(1,1,2,3,3,4,6,7)$	1	3	15120	3^3	3^5
108d	$(1,1,2,3,3,4,6,7)$	1	4	7560	3^3	3
108e	$(1,1,2,3,3,4,6,7)$	1	6	7560	3^3	3^3
108f	$(1,1,2,3,3,4,6,7)$	1	7	7560	3^3	3
109a	$(1,1,2,3,3,5,6,6)$	2	1	7560	3^7	3
109b	$(1,1,2,3,3,5,6,6)$	2	2	3780	3^7	3
109c	$(1,1,2,3,3,5,6,6)$	2	3	7560	3^7	3^3
109d	$(1,1,2,3,3,5,6,6)$	2	5	3780	3^7	$3 \cdot 19^2$
109e	$(1,1,2,3,3,5,6,6)$	2	6	7560	3^7	$2^6 \cdot 3^3$
110a	$(1,1,2,3,4,4,6,6)$	2	1	7560	3^3	$3 \cdot 37^2$
110b	$(1,1,2,3,4,4,6,6)$	2	2	3780	3^3	$3 \cdot 73^2$
110c	$(1,1,2,3,4,4,6,6)$	2	3	3780	3^3	3^7
110d	$(1,1,2,3,4,4,6,6)$	2	4	7560	3^3	$3 \cdot 19^2$
110e	$(1,1,2,3,4,4,6,6)$	2	6	7560	3^3	$2^6 \cdot 3^3$
111a	$(1,1,3,3,3,3,6,7)$	2	1	1260	3^3	3
111b	$(1,1,3,3,3,3,6,7)$	2	3	2520	3^3	3^5
111c	$(1,1,3,3,3,3,6,7)$	2	6	630	3^3	$3^3 \cdot 17^2$
111d	$(1,1,3,3,3,3,6,7)$	2	7	630	3^3	$3 \cdot 53^2$
112a	$(1,1,3,3,3,4,5,7)$	2	1	5040	3^3	3
112b	$(1,1,3,3,3,4,5,7)$	2	3	7560	3^3	3^5

	a	$w(3)$	a_0	mult.	$c_1 = 1$	c_2
112c	$(1,1,3,3,3,4,5,7)$	2	4	2520	3^3	$3 \cdot 37^2$
112d	$(1,1,3,3,3,4,5,7)$	2	5	2520	3^3	$2^6 \cdot 3$
112e	$(1,1,3,3,3,4,5,7)$	2	7	2520	3^3	$3 \cdot 53^2$
113a	$(1,1,3,3,3,4,6,6)$	1	1	2520	3^3	$2^6 \cdot 3$
113b	$(1,1,3,3,3,4,6,6)$	1	3	3780	3^3	3^5
113c	$(1,1,3,3,3,4,6,6)$	1	4	1260	3^3	3
113d	$(1,1,3,3,3,4,6,6)$	1	6	2520	3^3	3^3
114a	$(1,1,3,6,6,6,6,7)$	1	1	1260	3^3	3
114b	$(1,1,3,6,6,6,6,7)$	1	3	630	3^3	3^5
114c	$(1,1,3,6,6,6,6,7)$	1	6	2520	3^3	3^3
114d	$(1,1,3,6,6,6,6,7)$	1	7	630	3^3	3
115a	$*(1,1,4,4,6,6,7,7)$	0	1	1260	3^3	3
115b	$*(1,1,4,4,6,6,7,7)$	0	4	1260	3^3	3
115c	$*(1,1,4,4,6,6,7,7)$	0	6	1260	3^3	0
115d	$*(1,1,4,4,6,6,7,7)$	0	7	1260	3^3	3
116a	$(1,1,4,5,6,6,6,7)$	1	1	5040	3^3	3
116b	$(1,1,4,5,6,6,6,7)$	1	4	2520	3^3	$3 \cdot 17^2$
116c	$(1,1,4,5,6,6,6,7)$	1	5	2520	3^3	3
116d	$(1,1,4,5,6,6,6,7)$	1	6	7560	3^3	3^3
116e	$(1,1,4,5,6,6,6,7)$	1	7	2520	3^3	3
117a	$(1,1,4,6,6,6,6,6)$	2	1	252	$3^3 \cdot 17^2$	$3 \cdot 37^2$
117b	$(1,1,4,6,6,6,6,6)$	2	4	126	$3^3 \cdot 17^2$	$3 \cdot 19^2$
117c	$(1,1,4,6,6,6,6,6)$	2	6	630	$3^3 \cdot 17^2$	3^3
118a	$(1,2,3,3,3,3,4,8)$	2	1	1260	3^5	3
118b	$(1,2,3,3,3,3,4,8)$	2	2	1260	3^5	$3 \cdot 19^2$
118c	$(1,2,3,3,3,3,4,8)$	2	3	5040	3^5	$3^3 \cdot 17^2$
118d	$(1,2,3,3,3,3,4,8)$	2	4	1260	3^5	$3 \cdot 37^2$
118e	$(1,2,3,3,3,3,4,8)$	2	8	1260	3^5	$2^6 \cdot 3$
119a	$(1,2,3,3,3,3,6,6)$	1	1	630	3^5	3
119b	$(1,2,3,3,3,3,6,6)$	1	2	630	3^5	$3 \cdot 17^2$

	a	$w(3)$	a_0	mult.	$\mathbf{c_1 = 1}$	$\mathbf{c_2}$
119c	$(1,2,3,3,3,3,6,6)$	1	3	2520	3^5	3^3
119d	$(1,2,3,3,3,3,6,6)$	1	6	1260	3^5	3^3
120a	$(1,2,3,3,3,4,5,6)$	1	1	5040	3^5	3
120b	$(1,2,3,3,3,4,5,6)$	1	2	5040	3^5	$3 \cdot 17^2$
120c	$(1,2,3,3,3,4,5,6)$	1	3	15120	3^5	3^3
120d	$(1,2,3,3,3,4,5,6)$	1	4	5040	3^5	3
120e	$(1,2,3,3,3,4,5,6)$	1	5	5040	3^5	3
120f	$(1,2,3,3,3,4,5,6)$	1	6	5040	3^5	3^3
121a	$*(1,2,3,3,6,6,7,8)$	0	1	3780	0	3
121b	$*(1,2,3,3,6,6,7,8)$	0	2	3780	0	3
121c	$*(1,2,3,3,6,6,7,8)$	0	3	7560	0	3^3
121d	$*(1,2,3,3,6,6,7,8)$	0	6	7560	0	3^3
121e	$*(1,2,3,3,6,6,7,8)$	0	7	3780	0	3
121f	$*(1,2,3,3,6,6,7,8)$	0	8	3780	0	3
122a	$*(1,2,3,4,5,6,7,8)$	0	1	5040	0	3
122b	$*(1,2,3,4,5,6,7,8)$	0	2	5040	0	3
122c	$*(1,2,3,4,5,6,7,8)$	0	3	5040	0	3^3
122d	$*(1,2,3,4,5,6,7,8)$	0	4	5040	0	3
122e	$*(1,2,3,4,5,6,7,8)$	0	5	5040	0	3
122f	$*(1,2,3,4,5,6,7,8)$	0	6	5040	0	3^3
122g	$*(1,2,3,4,5,6,7,8)$	0	7	5040	0	3
122h	$*(1,2,3,4,5,6,7,8)$	0	8	5040	0	3
123a	$(1,2,3,6,6,6,6,6)$	2	1	252	3^5	$3 \cdot 19^2$
123b	$(1,2,3,6,6,6,6,6)$	2	2	252	3^5	$3 \cdot 37^2$
123c	$(1,2,3,6,6,6,6,6)$	2	3	252	3^5	3^3
123d	$(1,2,3,6,6,6,6,6)$	2	6	1260	3^5	$3^3 \cdot 17^2$
124a	$(1,3,3,3,3,3,3,8)$	3	1	42	3^9	$3 \cdot 73^2$
124b	$(1,3,3,3,3,3,3,8)$	3	3	252	3^9	$2^6 \cdot 3^3$
124c	$(1,3,3,3,3,3,3,8)$	3	8	42	3^9	$3 \cdot 17^2$
125a	$(1,3,3,3,3,3,4,7)$	3	1	84	$3^3 \cdot 5^6$	$3 \cdot 73^2$

	a	$w(3)$	a_0	mult.	$c_1 = 1$	c_2
125b	$(1,3,3,3,3,3,4,7)$	3	3	420	$3^3 \cdot 5^6$	3^9
125c	$(1,3,3,3,3,3,4,7)$	3	4	84	$3^3 \cdot 5^6$	$3 \cdot 73^2$
125d	$(1,3,3,3,3,3,4,7)$	3	7	84	$3^3 \cdot 5^6$	$3 \cdot 73^2$
126a	$*(1,3,3,3,6,6,6,8)$	0	1	420	0	3
126b	$*(1,3,3,3,6,6,6,8)$	0	3	1260	0	3^3
126c	$*(1,3,3,3,6,6,6,8)$	0	6	1260	0	3^3
126d	$*(1,3,3,3,6,6,6,8)$	0	8	420	0	3
127a	$*(1,3,3,4,6,6,6,7)$	0	1	840	3^3	3
127b	$*(1,3,3,4,6,6,6,7)$	0	3	1680	3^3	0
127c	$*(1,3,3,4,6,6,6,7)$	0	4	840	3^3	3
127d	$*(1,3,3,4,6,6,6,7)$	0	6	2520	3^3	3^3
127e	$*(1,3,3,4,6,6,6,7)$	0	7	840	3^3	3
128a	$(3,3,3,3,3,3,3,6)$	3	3	14	3^9	$2^6 \cdot 3^3$
128b	$(3,3,3,3,3,3,3,6)$	3	6	2	3^9	$3^3 \cdot 5^6$
129a	$*(3,3,3,3,6,6,6,6)$	0	3	35	0	3^3
129b	$*(3,3,3,3,6,6,6,6)$	0	6	35	0	3^3

(c) Let $m = 6$, $n = 4$, $p = 7$. The following table records the values of

$$p^{w(2)} \text{Norm}\left(1 - \frac{\mathfrak{J}(\mathbf{c}, \mathbf{a})}{p^2}\right)$$

for $c_1 = 1$ and $c_2 = (5, 1, 1, 1, 1, 1)$, with $\mathbf{a}$ running through a list of representatives of the isomorphism classes of twisted Fermat motives. As usual, we break the 24 isomorphism classes of Fermat motives into subclasses determined by the first coefficient, a_0. Supersingular motives are marked by an asterisk.

	a	$w(2)$	a_0	mult.	$c_1 = 1$	c_2
1a	$(1,3,5,5,5,5)$	1	1	10	1	3
1b	$(1,3,5,5,5,5)$	1	3	10	1	3^3
1c	$(1,3,5,5,5,5)$	1	5	40	1	$2^2 \cdot 3$
2a	$*(1,1,1,5,5,5)$	0	1	10	2^2	3
2b	$*(1,1,1,5,5,5)$	0	5	10	2^2	3

	a	$w(2)$	a_0	mult.	$c_1 = 1$	c_2
$3a$	$(1,4,4,5,5,5)$	1	1	20	$2^2 \cdot 3$	1
$3b$	$(1,4,4,5,5,5)$	1	4	40	$2^2 \cdot 3$	3^3
$3c$	$(1,4,4,5,5,5)$	1	5	60	$2^2 \cdot 3$	5^2
$4a$	$*(1,1,2,4,5,5)$	0	1	60	0	1
$4b$	$*(1,1,2,4,5,5)$	0	2	30	0	3
$4c$	$*(1,1,2,4,5,5)$	0	4	30	0	3
$4d$	$*(1,1,2,4,5,5)$	0	5	60	0	1
$5a$	$*(1,1,3,3,5,5)$	0	1	30	2^2	3
$5b$	$*(1,1,3,3,5,5)$	0	3	30	2^2	0
$5c$	$*(1,1,3,3,5,5)$	0	5	30	2^2	3
$6a$	$*(1,2,2,3,5,5)$	0	1	60	3	1
$6b$	$*(1,2,2,3,5,5)$	0	2	120	3	0
$6c$	$*(1,2,2,3,5,5)$	0	3	60	3	1
$6d$	$*(1,2,2,3,5,5)$	0	5	120	3	2^2
$7a$	$(1,1,1,1,1,1)$	2	1	2	13^2	$3 \cdot 5^2$
$8b$	$(1,1,1,1,4,4)$	1	1	20	3	1
$8c$	$(1,1,1,1,4,4)$	1	4	10	3	3^3
$9a$	$(1,1,1,2,3,4)$	1	1	120	3^3	2^4
$9b$	$(1,1,1,2,3,4)$	1	2	40	3^3	3
$9c$	$(1,1,1,2,3,4)$	1	3	40	3^3	1
$9d$	$(1,1,1,2,3,4)$	1	4	40	3^3	$2^2 \cdot 3$
$10a$	$(1,1,1,3,3,3)$	1	1	20	1	$2^2 \cdot 3$
$10b$	$(1,1,1,3,3,3)$	1	3	20	1	3^3
$11a$	$(1,1,2,2,2,4)$	1	1	40	2^4	3
$11b$	$(1,1,2,2,2,4)$	1	2	60	2^4	1
$11c$	$(1,1,2,2,2,4)$	1	4	20	2^4	5^2
$12a$	$(1,1,2,2,3,3)$	1	1	60	$2^2 \cdot 3$	5^2
$12b$	$(1,1,2,2,3,3)$	1	2	60	$2^2 \cdot 3$	3^3
$12c$	$(1,1,2,2,3,3)$	1	3	60	$2^2 \cdot 3$	2^4
$13a$	$*(1,1,4,4,4,4)$	0	1	10	1	3

	a	$w(2)$	a_0	mult.	$c_1 = 1$	c_2
13b	$*(1,1,4,4,4,4)$	0	4	20	1	1
14a	$(1,2,2,2,2,3)$	1	1	10	5^2	3^3
14b	$(1,2,2,2,2,3)$	1	2	40	5^2	2^4
14c	$(1,2,2,2,2,3)$	1	3	10	5^2	3
15a	$*(1,2,2,4,4,5)$	0	1	30	2^2	3
15b	$*(1,2,2,4,4,5)$	0	2	60	2^2	1
15c	$*(1,2,2,4,4,5)$	0	4	60	2^2	1
15d	$*(1,2,2,4,4,5)$	0	5	30	2^2	3
16a	$*(1,2,3,3,4,5)$	0	1	60	0	1
16b	$*(1,2,3,3,4,5)$	0	2	60	0	3
16c	$*(1,2,3,3,4,5)$	0	3	120	0	2^2
16d	$*(1,2,3,3,4,5)$	0	4	60	0	3
16e	$*(1,2,3,3,4,5)$	0	5	60	0	1
17a	$*(1,2,3,4,4,4)$	0	1	40	1	0
17b	$*(1,2,3,4,4,4)$	0	2	40	1	1
17c	$*(1,2,3,4,4,4)$	0	3	40	1	3
17d	$*(1,2,3,4,4,4)$	0	4	120	1	2^2
18a	$*(1,3,3,3,3,5)$	0	1	5	2^2	3
18b	$*(1,3,3,3,3,5)$	0	3	20	2^2	0
18c	$*(1,3,3,3,3,5)$	0	5	5	2^2	3
19a	$*(1,3,3,3,4,4)$	0	1	20	3	2^2
19b	$*(1,3,3,3,4,4)$	0	3	60	3	1
19c	$*(1,3,3,3,4,4)$	0	4	40	3	0
20a	$(2,2,2,2,2,2)$	1	2	2	3^3	3
21a	$*(2,2,2,4,4,4)$	0	2	10	0	3
21b	$*(2,2,2,4,4,4)$	0	4	10	0	3
22a	$*(2,2,3,3,4,4)$	0	2	30	2^2	1
22b	$*(2,2,3,3,4,4)$	0	3	30	2^2	0
22c	$*(2,2,3,3,4,4)$	0	4	30	2^2	1
23a	$*(2,3,3,3,3,4)$	0	2	5	0	3

	a	$w(2)$	a_0	mult.	$c_1 = 1$	c_2
23b	$*(2,3,3,3,3,4)$	0	3	20	0	2^2
23c	$*(2,3,3,3,3,4)$	0	4	5	0	3
24a	$*(3,3,3,3,3,3)$	0	3	1	2^2	0

B   How to compute the stable Picard number when m is prime

In this appendix we briefly describe the main ideas in Shioda's method and how they can be used to compute the stable combinatorial Picard number when m is prime and $p \equiv 1$ (mod m), yielding the formulas in Proposition 5.4. All of the ideas in this section are due to Shioda [Sh79a]. The main idea is to use the inductive structure to construct supersingular characters.

Fix m prime, $n = 2d$ even, and assume $p \equiv 1$ (mod m). Consider a character $\mathbf{a} = (a_0, a_1, \ldots, a_{n+1})$. Since the property of being supersingular is invariant under permutation of the entries of the vector $(a_0, a_1, \ldots, a_{n+1})$, we will work with $\mathbf{a}$ up to such permutation. This means that all we need to consider is the multiplicity vector $(x_1, x_2, \ldots, x_{m-1})$, where we set $x_i = x_i(\mathbf{a}) = \#\{a_j \mid a_j = i\}$. In fact, it will turn out to be convenient to augment this multiplicity vector by adding one last entry equal to one more than the length of $\mathbf{a}$, so we define

$$\mathbf{x}(\mathbf{a}) = (x_1(\mathbf{a}), x_2(\mathbf{a}), \ldots, x_{m-1}(\mathbf{a}); \|\mathbf{a}\| + 1).$$

If, for an integer z, we denote by $\{z\}$ the unique integer between 0 and $m-1$ which is congruent to z modulo m, then the definition of the length of a vector says that

$$\sum_{i=1}^{m-1} \{i x_i(\mathbf{a})\} = m(\|\mathbf{a}\| + 1). \qquad (*)$$

(This explains why we augment the vector by adding $\|\mathbf{a}\| + 1$ rather than $\|\mathbf{a}\|$, and it also gives a condition that any augmented multiplicity vector must satisfy.) Finally, we can recover the dimension n from the vector $\mathbf{x}(\mathbf{a})$ by the formula

$$\sum_{i=1}^{m-1} x_i(\mathbf{a}) = n + 2.$$

For example, if we let $m = 5$, $n = 6$, and $\mathbf{a} = (1, 1, 2, 2, 3, 3, 4, 4)$, the associated multiplicity vector is $\mathbf{x} = (2, 2, 2, 2; 4)$. One checks easily that the condition given by equation $(*)$ is satisfied. Of course, the same multiplicity vector is attached to every permutation of $\mathbf{a}$.

Given a multiplicity vector $\mathbf{x} = (x_1, x_2, \ldots, x_{m-1}; y)$, it is easy to compute how many characters $\mathbf{a}$ it represents: the number is simply

$$\#\mathbf{x} = \frac{(x_1 + x_2 + \cdots + x_{m-1})!}{x_1! x_2! \cdots x_{m-1}!} = \frac{(n+2)!}{x_1! x_2! \cdots x_{m-1}!}.$$

Of course, we really want to consider only those vectors where y is as determined by the relation implied by formula $(*)$, that is,

$$\sum_{i=1}^{m-1} \{i x_i\} = my. \tag{$\dagger$}$$

To compute the stable combinatorial Picard number, we need to count all the supersingular characters for a given m and n (and $p \equiv 1 \pmod{m}$). We will do this by first counting all the relevant multiplicity vectors, then considering their permutations. So we must begin by characterizing which multiplicity vectors correspond to supersingular characters. This is easy to do by using Proposition 3.8: since $p \equiv 1 \pmod{m}$, we see that $\mathbf{a}$ is supersingular if and only if we have $\|t\mathbf{a}\| = n/2$ for every $t \in (\mathbb{Z}/m\mathbb{Z})^\times$. Translating to multiplicity vectors, we see that a vector $(x_1, x_2, \ldots, x_{m-1}; y)$ corresponds to a supersingular character if and only if we have

$$\sum_{i=1}^{m-1} \{i t x_i\} = my \qquad \text{for all } t \in (\mathbb{Z}/m\mathbb{Z})^\times. \tag{$\ddagger$}$$

It is easy to show that if the system of equations $(\ddagger)$ holds and $n = x_1 + x_2 + \cdots + x_{m-1} - 2$ is the dimension, then $y = n/2 + 1$. Thus, we can characterize the multiplicity vectors corresponding to supersingular characters as the solutions of the system of diophantine equations $(\ddagger)$.

Following Shioda, let us denote the set of all solutions of the system by M_m and the set of solutions with $y = s$ by $M_m(s)$, so that $M_m(s)$ contains the multiplicity vectors that correspond to supersingular characters of dimension $2s - 2$. It is clear that M_m is an additive semigroup; more precisely, if $\mathbf{x} = (x_1, x_2, \ldots, x_{m-1}; y) \in M_m(s)$ and $\mathbf{x}' = (x_1', x_2', \ldots, x_{m-1}'; y') \in M_m(s')$, then $\mathbf{x} + \mathbf{x}' = (x_1 + x_1', x_2 + x_2', \ldots, x_{m-1} + x_{m-1}'; y + y')$ belongs to $M_m(s + s')$ (making this formula work is another reason for using $\|\mathbf{a}\| + 1$ instead of $\|\mathbf{a}\|$.)

The crucial fact, already noted by Ran [Ran81], is:

PROPOSITION B.1
If m is prime, M_m is generated by $M_m(1)$.

For a proof, see the main proposition in Koblitz and Ogus [KO79].

So we first need to determine $M_m(1)$, but this is easy: the vectors in $M_m(1)$ are those for which $x_i = x_{m-i} = 1$ for some i and $x_k = 0$ otherwise. They correspond to characters (of dimension zero) $\mathbf{a} = (i, m - i)$. Thus, the proposition translates to:

PROPOSITION B.2

Suppose m is prime and $p \equiv 1 \pmod{m}$. Up to permutation, any supersingular character of dimension $n = 2d$ is of the form

$$(b_1, m - b_1, b_2, m - b_2, \ldots, b_d, m - b_d).$$

Proof: Straight translation from the previous proposition. $\square$

To explain how to establish the formulas in Proposition 5.4, we do the case $n = 4$ in detail. The case $n = 6$ is analogous, and the method can clearly be used (with increasing complications, of course) for any n.

Assume $n = 4$, m is prime, and $p \equiv 1 \pmod{m}$. We want to determine the vectors in $M_m(3)$. By the proposition, these are of the form $\mathbf{x} = \mathbf{x}' + \mathbf{x}'' + \mathbf{x}'''$, where $\mathbf{x}', \mathbf{x}'', \mathbf{x}''' \in M_m(1)$. There are several cases to consider, corresponding to the additive partitions of 3:

$3 = 3$: If $\mathbf{x}' = \mathbf{x}'' = \mathbf{x}'''$, then we have $x_i = x_{m-i} = 3$ for some i and $x_j = 0$ otherwise. There are $(m-1)/2$ choices for i, and each corresponds to $6!/3!3! = 20$ characters.

$3 = 1 + 2$: If $\mathbf{x}' = \mathbf{x}'' \neq \mathbf{x}'''$, then we have $x_i = x_{m-i} = 2$ for some i, $x_j = x_{m-j} = 1$ for some $j \neq i$, and $x_k = 0$ otherwise. There are $\binom{(m-1)/2}{2} = (m-1)(m-3)/8$ choices for i and j, and each corresponds to $6!/2!2! = 180$ characters.

$3 = 1 + 1 + 1$: If all three summands are distinct, one has $x_i = x_{m-i} = 1$ for three distinct i, and $x_j = 0$ otherwise. As before, there are $\binom{(m-1)/2}{3} = (m-1)(m-3)(m-5)/48$ choices for the three indices, and each multiplicity vector corresponds to $6! = 720$ characters.

Putting all of these together, we get that there are

$$20\frac{m-1}{2} + 180\frac{(m-1)(m-3)}{4} + 720\frac{(m-1)(m-3)(m-5)}{48}$$

$$= 5(m-1)(3m^2 - 15m + 20)$$

supersingular characters, which proves the formula for the case $n = 4$ in Proposition 5.4.

The case $n = 6$ is similar, except that the number of cases that need to be considered increases.

We conclude with a quick sketch of Shioda's approach to the Hodge conjecture for complex Fermat hypersurfaces (and hence to the Tate conjecture for Fermat hypersurfaces over finite fields when $p \equiv 1 \pmod{m}$). More details can be found in [Sh79a, Sh79b, Sh83b]. Let $n = 2d$. If m is prime, we have seen that M_m is generated by $M_m(1)$. For more general m, Shioda calls a vector $\mathbf{x} \in M_m(d+1)$ *decomposable* if $\mathbf{x} = \mathbf{x}' + \mathbf{x}''$ with $\mathbf{x}', \mathbf{x}'' \in M_m$, *quasi-decomposable* if there exists $\tilde{\mathbf{x}} \in M_m(s)$ with $s \le 2$ such that $\mathbf{x} + \tilde{\mathbf{x}} = \mathbf{x}' + \mathbf{x}''$ with $\mathbf{x}', \mathbf{x}'' \in M_m$, and *semi-decomposable* there exist vectors $(x_1', \ldots, x_{m-1}')$ and $(x_1'', \ldots, x_{m-1}'')$ such that we have $x_i = x_i' + x_i''$ for all i, and the x_i' and x_i'' satisfy the conditions

$$\sum i x_i' \equiv \sum i x_i'' \equiv 0 \pmod{m} \quad \text{and} \quad \sum x_i' = \sum x_i'' = 3.$$

If $\mathbf{x}$ is decomposable or quasi-decomposable, it corresponds to supersingular characters that can be obtained via the inductive structure, up to permutation, from supersingular characters of smaller dimension. For example, Proposition B.1 says that when m is prime every $\mathbf{x}$ is decomposable, and that we can take $\mathbf{x}'$ and $\mathbf{x}''$ to be in $M_m(1)$. (The semi-decomposable case is exceptional, and it is only relevant when $\mathbf{x} \in M_m(3)$. In this case, one is relating supersingular characters of dimension 4 to characters of dimension 1.)

Since Shioda can realize the inductive structure at the level of cohomology, he can use it to generate cohomology classes inductively. This yields a proof of the Tate conjecture *provided one knows that every supersingular character is either decomposable, quasi-decomposable, or semi-decomposable.* This is the condition P_m^n mentioned above. Shioda shows that if P_m^n holds for a given m and n, then so does Tate conjecture (for the case $p \equiv 1 \pmod{m}$; the general case requires a modified—and harder to use—form of the condition). This is the method used in [Sh79a, Sh79b] to prove the Tate conjecture in the cases specified in Proposition 5.17.

Bibliography

[AS83]   Aoki, N., and Shioda, T., "Generators of the Néron-Severi group of a Fermat surface", in *Arithmetic and Geometry*, Volume I, Progress in Math., Vol. **35**, Birkhäuser Boston 1983, pp. 1–12.

[AM77]   Artin, M., and Mazur, B., "Formal groups arising from algebraic varieties", *An. Sci. Ec. Norm. Sup.*, 4ème série **10** (1977), pp. 87–132.

[BN78]   Bayer, P., and Neukirch, J., "On values of zeta-functions and ℓ-adic Euler characteristics", *Invent. Math.*, **50** (1978), pp. 35–64.

[BK90]   Bloch, S., and Kato, K., "L-functions and Tamagawa numbers of motives", in *The Grothendieck Festschrift*, Vol I, Progress in Math., Vol. **88**, Birkhäuser Boston 1990, pp. 333–400.

[DH35]   Davenport, H., and Hasse, H., "Die Nullstellen der Kongruenz–Zetafunktionnen in gewissen zyklischen Fällen", *J. Reine Angew. Math.*, **172** (1935), pp. 151–182.

[De73]   Deligne, P., "Cohomologie des intersections complètes", Exp. XIX, *SGA 7II*, Lecture Notes in Mathematics **340**, Springer-Verlag 1973, pp. 401–428.

[Ek84]   Ekedahl, T., "On the multiplicative properties of the de Rham–Witt complex", *Archiv für Mathematik*, **22** (1984), pp. 185–239.

[Et88]   Etesse, J.-Y., "Rationalitié et valeurs de fonctions L en cohomologie cristalline", *Ann. Inst. Fourier*, **38**(4) (1988), pp. 33–92.

[GY92]   Gouvêa, F.Q., and Yui, N., "Brauer numbers of twisted Fermat motives", Preprint 1992, New York Number Theory Seminar (to appear).

[GGY]   Goto, Y., Gouvêa, F.Q., and N. Yui, "Arithmetic of weighted K3-surfaces over number fields", in preparation.

[Ha92]   Harrison, M., "On the conjecture of Bloch–Kato for Grosencharacters over $\mathbb{Q}(i)$", Ph.D. Thesis, University of Cambridge (1992).

[Ih86] Ihara, Y., "Profinite braid groups, Galois representations and complex multiplications", *Ann. of Math.*, **123** (1986), pp. 43–106.

[Iw75] Iwasawa, K., "A note on Jacobi sums", *Symposia Math.*, **15** (1975), pp. 447–459.

[IR83] Illusie, L., and Raynaud, M., "Les suites spectrales associeés au complexe de de Rham-Witt", *Publ. Math. IHES*, **57** (1983), pp. 73–212.

[Ko75] Koblitz, N., "*P*-adic variation of the zeta-function over families of varieties defined over finite fields", *Compositio Math.*, **31** (1975), pp. 119–219.

[KO79] Koblitz, N., and Ogus, A., "Algebraicity of some products of values of the Γ function", in *Automorphic Forms, Representations, and L-functions*, ed. by A. Borel and W. Casselman, Proceedings of Symposia in Pure Mathematics, Vol. **33**, part 2, American Mathematical Society 1979, pp. 343–345.

[Li84] Lichtenbaum, S., "Values of zeta-functions at non-negative integers", in *Number Theory*, Lecture Notes in Mathematics **1068**, Springer-Verlag 1984, pp. 129–138.

[Li87] Lichtenbaum, S., "The construction of weight-two arithmetic cohomology" , *Invent. Math.*, **88** (1987), pp. 183–215.

[Li90] Lichtenbaum, S., "New results on weight-two motivic cohomology", in *Grothendieck Festschrift*, Vol. III, Progress in Mathematics, **88**, Birkhaüser Boston 1990, pp. 35–56.

[Ma72] Mazur, B. "Frobenius and the Hodge filtration", *Bull. Amer. Math. Soc.*, **78** (1972), pp. 653–667.

[Mik87] Miki, H. "On the ℓ-adic expansion of certain Gauss sums and its applications", *Advanced Studies in Pure Math.*, **12** (1987), pp. 87–118.

[Mil75] Milne, J. S., "On a conjecture of Artin and Tate", *Ann. Math.*, **102** (1975), pp. 517–533.

[Mil86] Milne, J. S., "Values of zeta-functions of varieties over finite fields", *Amer. J. Math.*, **108** (1986), pp. 297–360.

[Mil88] Milne, J. S., "Motivic cohomology and values of zeta-functions", *Compositio Math.*, **68** (1988), pp. 59–102.

[Pa83] Parshin, A. V., "Chern classes, adeles and values of zeta–functions", *J. Reine Angew. Math.*, **341** (1983), pp. 174–192.

[PS91] Pinch, R., and Swinnerton-Dyer, H., "Arithmetic of diagonal quartic surfaces I", in *L-functions in Arithmetic*, Proceedings of the 1989 Durham Symposium, J. H. Coates and M. J. Taylor (eds.), Cambridge University Press 1991, pp. 317–338.

[Ran81] Ran, Z., "Cycles on Fermat hypersurfaces", *Compositio Math.*, **42** (1981), pp. 121–142.

[Sa89] Saito, S., "Arithmetic theory of arithmetic surfaces", *Ann. Math.*, **129** (1989), pp. 547–589.

[Sc82] Schneider, P., "On the values of the zeta-function of a variety over a finite field", *Compositio Math.*, **46** (1982), pp. 133–143.

[So90] Schoen, C., "Cyclic covers of $\mathbb{P}^\nu$ branched along $\nu + 2$ hyperplanes and the generalized Hodge conjecture for certain Abelian varieties", in *Lecture Notes in Mathematics* **1399**, Springer-Verlag 1990, pp. 137–154.

[SK79] Shioda, T., and Katsura, T., "On Fermat varieties", *Tohoku J. Math.*, **31** (1979), pp. 97–115.

[Sh79a] Shioda, T., "The Hodge conjecture for Fermat varieties", *Math. Ann.*, **245** (1979), pp. 175–184.

[Sh79b] Shioda, T., "The Hodge conjecture and the Tate conjecture for Fermat varieties", *Proc. Japan Academy*, **55**, Ser. A., No. 3 (1979), pp. 111–114.

[Sh81] Shioda, T., "Algebraic cycles on abelian varieties of Fermat type", *Math. Ann.*, **258** (1981), pp. 65–80.

[Sh82a] Shioda, T., "On the Picard number of a Fermat surface", *J. Fac. Sci. Univ. Tokyo*, Sec. IA **28**, No. 3 (1982), pp. 725–734.

[Sh82b] Shioda, T., "Geometry of Fermat varieties", in *Number Theory Related to Fermat's Last Theorem*, Progress in Math., **26** (1982), pp. 45–56.

[Sh83a] Shioda, T., "Algebraic cycles on certain hypersurfaces", in *Algebraic Geometry*, Proceedings Tokyo/Kyoto 1982, M. Raynaud and T. Shioda (eds.), Lecture Notes in Mathematics **1016**, Springer-Verlag 1983, pp. 271–294.

[Sh83b] Shioda, T., "What is known about the Hodge conjecture?", *Advanced Studies in Pure Math.*, Algebraic Varieties and Analytic Varieties, **1** (1983), pp. 55–68.

[Sh87]   Shioda, T., "Some observations on Jacobi sums", *Advanced Studies in Pure Math.*, Galois Representations and Arithmetic Algebraic Geometry, **12** (1987), pp. 119–135.

[So84]   Soulé, C., "Groupes de Chow et K–theorie de varieties sur un corps fini", *Math. Ann.*, **266** (1984), pp. 317–345.

[SY88]   Suwa, N., and Yui, N., "Arithmetic of Fermat Varieties I: Fermat motives and p-adic cohomologies", MSRI Berkeley Preprint 1988.

[SY89]   Suwa, N., and Yui, N., "Jacobi sums, Fermat motives, and the Artin–Tate formula", *C. R. Math. Rep. Acad. Sci. Canada*, **11**, (1989), pp. 183–188.

[Su91a] Suwa, N., "Fermat Motives and the Artin-Tate Formula. I", *Proc. Japan Academy*, **67** (1991), pp. 104–107.

[Su91b] Suwa, N., "Fermat Motives and the Artin-Tate Formula. II", *Proc. Japan Academy*, **67** (1991), pp. 135–138.

[Su93]   Suwa, N., "Hodge-Witt cohomology of complete intersections", *J. Math. Soc. Japan*, **45** (1993), pp. 295–300.

[Su]   Suwa, N., "Some remarks on the Artin-Tate formula for diagonal hypersurfaces", preprint.

[Ta65]   Tate, J., "Algebraic cycles and poles of zeta functions", in *Arithmetical Algebraic Geometry*, ed. by O. F. G. Schilling, Harper and Row, New York 1965, pp. 93–110.

[Ta68]   Tate, J., "On a conjecture of Birch and Swinnerton-Dyer and a geometric analogue", in *Dix Exposes sur la Cohomologie des Schemas*, North–Holland, Amsterdam 1968, pp. 189–214.

[TS]   Toki, K. and Sasaki, R., "Fermat Varieties of Hodge–Witt Type", preprint.

[We49]   Weil, A., "Numbers of solutions of equations in finite fields", *Bull. Amer. Math. Soc.*, **55** (1949), pp. 497–508.

[We52]   Weil, A., "Jacobi sums as Grossencharaktere", *Trans. Amer. Math. Soc.*, **74** (1952), pp. 487–495.

[Yu91]   Yui, N., "Special values of zeta-functions of Fermat varieties over finite fields", *Number Theory, New York Seminars*, Springer–Verlag 1991, pp. 251–275.

[Yu94]   Yui, N., "On the norms of algebraic numbers associated to Jacobi sums", *J. Number Theory*, **47** (1994), pp. 106–129.

Index

slopes, 26, 27, 32

strongly supersingular, 33, 36, 76, 85

supersingular, 3, 33, 36, 40, 45, 47, 64, 65, 76, 79, 102

Tate conjecture, 1, 4, 5, 9, 52, 55, 56, 81, 87

twist

    extreme, 5, 9, 51, 55, 76, 77, 83, 84, 86, 89

    very mild, 51, 52

twisted Fermat motives, 3, 29–30, 98–102

twisted Jacobi sums, 11–14

twisting vectors, 2, 12

    equivalent, 3, 12

very mild twist, 51, 52

zeta-function, 22–23

    special value, 23, 59–60, 76, 79, 81, 87

This book was typeset using LaTeX with the AMS-LaTeX macros plus a number of other add-on packages. The main text font is Computer Modern, designed by D.E. Knuth. For mathematics, fonts included Knuth's Computer Modern, the AMS fonts, and the Euler mathematics fonts. The book was printed from camera-ready copy provided by the authors.

Printed in the United States
By Bookmasters